John Newman

Metallic Structures

Corrosion and Fouling, and their Prevention

John Newman

Metallic Structures
Corrosion and Fouling, and their Prevention

ISBN/EAN: 9783743444171

Printed in Europe, USA, Canada, Australia, Japan

Cover: Foto ©berggeist007 / pixelio.de

More available books at **www.hansebooks.com**

METALLIC STRUCTURES:

CORROSION AND FOULING, AND

THEIR PREVENTION

A Practical Aid-Book

TO THE SAFETY OF WORKS IN IRON AND STEEL,
AND OF SHIPS; AND TO THE SELECTION
OF PAINTS FOR THEM.

By JOHN NEWMAN

ASSOC. M. INST. C.E., F. IMPL. INST.

AUTHOR OF

EARTHWORK SLIPS AND SUBSIDENCES'; 'NOTES ON CONCRETE AND WORKS IN CONCRETE'
'SCAMPING TRICKS AND ODD KNOWLEDGE OCCASIONALLY PRACTISED UPON PUBLIC WORKS'
'NOTES ON CYLINDER BRIDGE PIERS, AND THE WELL SYSTEM OF FOUNDATIONS'
'FOUNDATIONS IN SAND' (LECTURE, ROYAL ENGINEERS' INSTITUTE, CHATHAM)
ETC. ETC. ETC.

London:
E. & F. N. SPON, 125 STRAND
New York:
SPON & CHAMBERLAIN, 12 CORTLANDT STREET
1896

PREFACE.

THIS book has been written, as it is believed none has appeared for a long period, to help those who may be intrusted with the design, erection, or maintenance of any kind of iron or steel structure, and to call to remembrance the salient points requiring attention, in order to prevent or arrest corrosion and fouling.

The author, having had his attention specially directed, during many years, to the subject of corrosion and fouling and their prevention, has placed his experiences on record, and that of others in all parts of the world, to whom he offers his acknowledgments, and also to the Editors of the various Technical Journals mentioned.

Chemical analyses of the composition of air, water, iron, and steel of various kinds are only referred to to illustrate the text where necessary; however, no one can more appreciate the incalculable value of chemical analysis than the author, and the great help and guide it is to the engineer; but analyses of air, rain, water, iron, and steel, are to be found in the various able and exhaustive works which relate to them, and therefore are not repeated; the aim here being to present, as concisely as possible, information which may be useful to engineers, constructors, architects, students, owners, and those having the care of iron or steel structures and ships, and to indicate some preventive and remedial measures that can be adopted for their preservation from corrosion and fouling.

It should always be remembered that the preservation of a metallic structure is only second in importance to that of its design and erection, and, unless corrosion and fouling are prevented or repressed, the time must come when the original strength will be so impaired as to be dangerous.

Fouling is distinct in many of its features from that of corrosion, it has, therefore, been treated in Part II. of this book, and much information will be found in it relating to the fouling and corrosion of ships, pile, bridge, promenade, and landing piers, pontoons, and similar works of construction, and anti-corrosive, and anti-corrosive and anti-fouling paints and compositions: products of very great influence in the preservation and protection from corrosion and fouling of metallic structures, whether submerged or unsubmerged.

The causes of fouling and corrosion in sea water are briefly examined, also the qualities required in an anti-corrosive and anti-fouling paint, and in other coatings used with the object of preventing the fouling and corrosion of ships and any submerged or partly submerged work, and corrosion in all.

The subject being of a comprehensive nature, the book is not intended to be an exhaustive treatise, but a kind of *vade mecum* written to supply concise practical information on a subject of much importance to all having to do with the design, construction, or maintenance of metallic structures, whether floating or fixed.

J. N.

LONDON: *November* 19, 1895.

CONTENTS.

PART I.

THE SAFETY AND PRESERVATION OF IRON AND STEEL STRUCTURES.

PART II.

THE PREVENTION OF FOULING AND CORROSION IN SUBMERGED STRUCTURES AND SHIPS.

CORROSION AND ITS PREVENTION.

PART I.

THE SAFETY AND PRESERVATION OF IRON AND STEEL STRUCTURES.

CHAPTER I.

THE IMPORTANCE OF THE EFFICIENT MAINTENANCE OF METALLIC STRUCTURES AND THE PREVENTION OF CORROSION.

IN nature there is no stability or permanence, for decay and reconstruction are everywhere to be seen, and nothing is known which does not undergo a change of state when subject to the action of intense heat, and most substances are more or less affected by water. Although no marked alteration of form or appearance may occur or be appreciable in the initiatory stages, a continual change proceeds in almost every material used in construction from the time it is exposed to any of the elements, or when in contact with many other substances, for the power of decomposition is exercised by different substances on others. The combustion of decay has commenced and cannot be entirely prevented, but it may be much lessened, and, in some cases, completely controlled. The process of oxidation has become visible on metals when they have become tarnished, but a metamorphic condition will have occurred before its effects are apparent, and under the surface it may be proceeding rapidly and spreading around, such decomposing action frequently having nuclei from which its power springs.

B

The maintenance and present condition of iron and steel used in various forms in engineering and building construction are likely to demand increasing attention, as either from fatigue, vibration, corrosion, decomposition, or general deterioration, the metal may have become either changed in its characteristics, or be so impaired as to be no longer the same metal as when erected, or of sufficient dimensions to sustain the load it originally was well able to support.

The number of metallic structures requiring either strengthening or renewing must increase, and, in thoroughly opened-up countries, it is probable their restoration will soon be an important branch of engineering science. It is one in which special skill is necessary, not only to ascertain the real condition of a structure, but also to arrest any elements of decay, and to restore it to its original strength. Metallic railway and road bridges, and public buildings of importance have more or less continued attention bestowed upon them, but the care is usually, and not unnaturally, commensurate with the size of the bridge or building, and its public position and importance, notwithstanding that in the smaller structures the surfaces may be more exposed, and corrosion greater in comparison with the sectional area of the metal. Private warehouses and buildings are usually left either to a tenant to keep in repair, which generally means a coat or two of cheap oil paint every few years, or are under the supervision of some local builder whose knowledge of the circumstances which cause corrosion or depreciation is not, to say the least, too profound.

Structures and buildings supported by iron or steel columns and pillars should be occasionally examined by an expert engineer, for in many cases, although the ironwork may be well designed so far as regards strength to carry any load, and with the view to easy erection, the instances are comparatively few in which their preservation from corrosion has been specially considered. However, metallic bridges will have to be rebuilt unless care is regularly taken to preserve all their parts; for corrosion, if allowed to proceed, will, sooner or later, have increased to such an extent that, apart from the question of the fatigue of the metal and additional requirements of traffic, the reduction of area will be so considerable as to demand attention. Mr. Ewing Matheson in his paper ' Steel for Structures,' * wrote, " The preservation of iron from rust is not in this country sufficiently considered." Steel being now so much used, and the sectional area reduced in many instances as compared with iron, the preservation of the metal in its original section is of even more importance than formerly.

Corrosion arises, and is promoted or accelerated by chemical, voltaic, and mechanical action. It therefore varies, for almost every metal is differently affected, even by mere mechanical action, for it is less on the surface of hard metals than soft, and the resistance to penetration by vapour or moisture is also augmented. Chemical and voltaic action will vary according to the composition and homogeneousness of the metal. This indicates the direction in which an explanation may be sought for any exceptional and peculiar corrosion, and may lead to the cause of it being discovered.

Metals, unlike Portland cement, which to a certain limiting period increases, or should increase, in strength with age, suffer a diminution of strength, however slow it may be, almost from the time they are used. To attempt to prevent corrosion without knowing the cause of such action can only by chance be successful, for the rapidity, and therefore the power of corrosive influences, depends upon the conditions and circumstances in which the metal is placed. Sir B. Baker has concisely declared that " it is the deviation from the average which really is so important in the design of engineering works." In a few instances it has happened that the reports of the most eminent professors of chemistry and the results obtained in engineering practice have not agreed, and yet undoubtedly *both* have been correct. The conditions under which the substances or liquids have been used has been the cause of the dissimilarity. In the

* 'Minutes of Proceedings, Inst. C.E.,' vol. lxix.

laboratory the examination is conducted with the greatest minuteness, and time is allowed for important action, and attention is specially directed to discover everything that can be detected regarding the information desired. If all the conditions are not clearly stated under which a substance or liquid is to be used, a report will most probably be either too favourable or unfavourable, although perfectly correct so far as probable results under laboratorial circumstances are concerned. It is necessary for an engineer to fully explain to an analytical chemist the corrosive influences to which any substance will be exposed, and the circumstances in which it will be used.

Somewhat amusing instances have occurred because of the want of such concerted action, for it was said Portland cement was most injuriously affected by sea water, and so it may be in certain experiments, and it may be said so would any other substance, but good Portland cement, properly mixed and deposited, can be thoroughly relied upon under the conditions in which it is used in concrete structures in sea water, and has been most successfully and economically for many years; however, the elaborate disintegration and action caused by minute analysis would probably generally cause an adverse view to be formed of its adaptability for such purposes. *The different circumstances* in which the material was used being alone responsible for the dissimilarity of action, for in analysis the Portland cement was minutely subdivided, whereas in a structure only the surface of a mass is exposed.

Similarly, sea water for watering streets was reported upon unfavourably, but it is adopted, and with success, either for sewers or streets at several fashionable watering places, where it is almost a necessity of the town's existence that everything should be so arranged as to attract and be approved. The apparent discrepancy in the reports of the analytical chemists and the results found by engineers was caused by the different circumstances of the tests and the actual use in roads, &c. In the case of road-watering, the action of the sea water was merely superficial, and by reason of the road being hardened by its binding action and caked, there was less dust to bo blown about; but macadamised roads or paving setts, if frequently watered by sea water, became slippery. When used for sewer flushing, if done quickly, and with a sufficient quantity of sea water, and if time is not given for fermentative action to develop, nothing unpleasant or injurious occurs. The circumstances of practical use modified the expected results determined by experiments in a laboratory, and yet while apparently impugning them, confirmed their accuracy. From these examples it can be readily judged, care should be taken to give full information to an analytical chemist in any special case in which the probable intensity of corrosive influences is desired to be known.

Although mechanical tests are usually of more value to the engineer than chemical analyses, still the latter are always valuable, and so far as corrosion is concerned its probable progress cannot be determined except from the probable chemical action.

It is gradually being considered a matter of vital importance in all countries to adopt some means of ascertaining the extent to which a metallic bridge or structure has deteriorated or is deteriorating; and the paramount necessity of preserving the full powers of the metal is acknowledged in order that the required strength may be retained for the longest possible period. Corrosion and general deterioration, if allowed to proceed unchecked, must culminate in failure. In order to be able to form a correct judgment of the magnitude of the corrosion of any structure, it is necessary to give instructions in case a personal inspection cannot be made, so that correct data are collected. Some of the chief points requiring answers are :—

1. Situation.
2. Date of erection of the bridge.
3. Name and address of the bridge builder.
4. Materials used in the various parts.

5. State if any alteration in the character of the neigh-bourhood since erection, or if anything has occurred since erection or last inspection likely to produce a more corrosive atmosphere.

6. Whether any structural alterations have been made. If so, state them.

7. If any alteration in the load or traffic since erection, whether increased weight of engine or rolling stock, speed or traffic.

8. Thickness of the timber bearers between rail or chair base and girder flange, and whether it has been lessened or increased.

9. Permanent deflection.

10. Deflections on loading tests.

11. State whether any parts have been renewed.

12. Exact distance apart of the top and bottom flanges.

13. Length of the lattice bars or diagonals.

14. If any lateral distortion in the girders.

15. Whether the vibration is excessive or not.

16. The state of the joints and the riveting.

17. The state of the paint.

18. Whether all the parts are free from corrosion, and mention those corroded, if any.

19. Quantity and weight of rust, if any, taken from structure. Top and bottom flanges, web, bars, bracing and wind-ties, flooring to be kept separate.

20. State if anything is observed influencing the corrosion or decay of the structure, and is not previously mentioned.

A few of the preceding questions are only applicable to bridges, but those for metallic structures generally can be easily separated.

The amount of the rust removed, if carefully collected under the superintendence of an expert, would periodically give an idea as to the relative rate of corrosion, and any exceptional initial quantity or increase would show that the durability of the structure was seriously affected; also, if the amount of rust taken from each part was compared with that removed from the same part at the previous inspection, and the rate of corrosion of each part computed per unit of area, some indication, it could hardly be exact, would be obtained of the places especially liable to corrosion.

In considering the data it is necessary to bear in mind that much of the strength of metallic bridges depends upon the metal retaining its elasticity, original sectional area, freedom from corrosion and buckling, absence of any loose or weak joints, or change in the fibre or the structure of the metal. If a bridge cannot be tested by loading, consequent upon the exigencies of the traffic, much can be gathered as to its condition by examination by an expert engineer; for the permanent way men can easily detect rottenness in any timber, but cannot be expected to know much about the strains brought upon girders or their maximum safe deflection under a passing load, or whether a permanent set has taken place, which, if it be permitted to continue, will probably ultimately be disastrous.

The importance of frequent periodical examination of existing metallic bridges and structures cannot be over-estimated, and also their freedom from corrosion when first properly coated with a really preservative and anti-corrosive paint. In the United States of America many engineers favour periodical inspection by a skilled engineer who is an expert in bridge construction, and that the inspection should be compelled by legislative authority although not undertaken by it, but signed and certified reports sent in to be examined by a committee of experts, as occasionally bridges have proved by failure that they were " all wrong," when reported by foremen to be " all right."

The number of old bridges, which is gradually being lessened, receiving the necessary attention, subject to increased loads, additional traffic, and at an accelerated speed, is not inconsiderable, and it is the smaller bridges rather than those of considerable span that perhaps more especially require inspection periodically, as they can hardly receive the same amount of care as any important structures, and yet the consequences of failure may be very serious indeed. The examination should be made by experts in metallic construction and repairs, or those trained to make such reports, and the time may come when there will be specially appointed engineers to make such examinations and reports, suggest precautionary and remedial measures, and carry them into execution.

Not many lines of a specification are generally reserved for protection against corrosion, and some elaborate specifications in all other respects may refer to painting in something like the following few words, " all iron surfaces to be painted with two coats of metallic paint and oil, and with an additional coat of lead and oil when the structure has been erected, the time of painting the last coat to be determined by the engineer." Thus the preservation of the metal from corrosion is only indirectly referred to, the covering of the metal with some substance being alone mentioned. Such a specification may be regarded as one which simply cares little for the maintenance of the original strength of a structure so long as it is erected.

In the case of any metallic structure which can only be occasionally inspected by an expert engineer, or when it has to be erected abroad, special provision should be made against corrosive influences.

Although calculations of the strains, alterations from the original design during construction, inferior material and workmanship, increased rolling load, the greater speed of the trains, unequal subsidence of the piers or abutments, or other special influences, are all important in their bearing on the serviceable life of a metallic bridge, that of proper maintenance is equally so, for if it be neglected, the strength of a structure is reduced, and cannot be restored except by additions or alterations, such as extra flange plates, stiffeners, web plates or bars, cross girders, and bracing. The metal may be said to be deteriorating no matter how slowly, from change of stress, vibration, and corrosion, and the chief aim is to reduce these deteriorating influences to a minimum.

The determination of the gradual encroachment of molecular changes in metal due to frequent alteration of strain is not fully known, but it is accepted that any such alteration must affect corrosion. There are phenomena that will probably be discovered in the behaviour of iron and steel which require time to determine ; for instance, whether cast iron especially, and wrought iron and steel in a less degree, which has been undisturbed from its original position for a long series of years, becomes crystallised or not so as to be almost brittle on being disturbed, or broken with a light blow. Any alteration in the composition or texture of metals will affect the intensity of the corrosion. The metal in a bridge is most probably, if not always, more or less deteriorating from the effects of variable strain, oxidation, vibration, and even, it is said, from a structure being fixed and strained more either from expansion or contraction on one side than the other.

It seems to be generally agreed that there are metallic bridges in use in various parts of the world which were not calculated to be strained as they are. The additional stress may be considered to arise from increase of load, speed, traffic, fatigue of the metal, loosening of the parts, the rails being fixed nearer the girders and without so thick an elastic medium as timber, and corrosion consequent upon inattention to the painting and its maintenance so as to prevent rust, or some of them; the influence of corrosion being all-important as being not only superficial, but also as reducing the sectional area of the metal, loosening the parts, and therefore augmenting the strains, or, it may be, altering their nature, particularly in bridges of small span.

The special danger of corrosion is when it from various causes produces exceptional *local weakness* or deterioration till the metal at such a point is strained beyond its limit of elasticity, then, although other portions of a metallic struc- ture may not be nearly so severely strained, the whole may fail, owing to want of structural stiffness, which it is equally as important to regard as mere strength. Some parts requiring special attention are the girder beds, bed and wall plates, the riveting, all junctions of parts, joints, and any place where water or dirt can accumulate, and places where different metals or materials meet, such as wrought iron with steel or cast iron, and any metal with timber; also, after years of service, the platforms of bridges, the abutments, parapets and wingwalls, whether of brickwork, masonry or concrete, may become in an unsatisfactory condition and aid in forming receptacles for the promotion of corrosion.

It has been said the causes of the deterioration of iron bridges are : (1) faulty design ; (2) faulty execution ; and (3) careless maintenance. In some countries the State regulations require : (1) a thorough test before opening for traffic; (2) the structure to be kept in good condition, which is understood to mean periodical inspection at intervals not greater than one year, annual cleaning, examination of all parts, removal of rust, and preservation by paint ; (3) a record kept of the strains in the structure ; (4) a repetition of the original tests at stated periods.

The simple rolling load is not always commended, because the platform and permanent way may be so made as to alter the distribution of the strains, and a sudden application of load or a dynamic test has therefore been suggested.

The formation of rust, it is considered, must be particularly guarded against by good painting, and by making all the parts easily accessible.

The magnitude of the deflection, although perhaps indicating the general condition of a girder, is regarded as inconclusive, not only because it has been observed that in many cases the centre of a girder has risen, but in those of comparatively large span the changes caused by temperature are important, and alter the form of a girder to an appreciable extent.

11

CHAPTER II.

SOME CAUSES OF CORROSION DESCRIBED.

In considering the various causes and influences that in some way or other produce corrosion, it is well to remember that probably in the whole range of the science and art of construction there is nothing more difficult to contend against than the decay of materials, or anything requiring so much constant care and diversified treatment, consequent upon the different nature of the materials, the various objects to which they are applied, and the changeableness of the influences that cause deterioration and decomposition. Almost from the moment of their manufacture, the metallic portions of a structure may be said to be subject to the combustion of decay which, it is decreed, must sooner or later overtake them; the aim of the engineer is to reduce to a minimum every deteriorating influence. Strictly speaking, submerged structures would be considered to be those constantly immersed; however, there are but few structures which are not occasionally partly covered with either rain, fresh or salt water, and therefore submerged and unsubmerged work, so far as corrosion is concerned, is almost a distinction without a difference, except that metals constantly immersed are not subject to such severe corrosive influences as those alternately wet and dry.

Metal having signs of corrosion or even the appearance of being "weathered," is in a condition of deterioration, and cannot be considered as equal in strength to the same iron fresh from the rolls or moulds; and as a rule the nearer rust is to the surface of iron the more oxide of iron there is in it.

Experiments have shown that corrosion, not exteriorly aided by galvanic action, on steel plates increases progressively. Those of Mr. Andrews, F.R.S., showed it to be 50 per cent. more the second year compared with the first.

Formerly it was sometimes thought that rust on the surface of iron acted as an efficient protection, but when it is on a plate it causes the metal beneath it to corrode under certain conditions, and any film that may be regarded as a protective covering is not of uniform thickness or character and is liable to peel and fall, for the rust on iron in the form of hydrated peroxides does not adhere to the surface upon which it is formed, but continues to accumulate until at length the flakes of rust become too heavy to be held together. It should not be considered a protection, but as the opposite, for although it acts temporarily as a shield to some extent upon the metal on which it has formed, it has a corrosive action of its own. If it were an effectual protection, corrosion would be arrested, whereas everywhere there is evidence that corrosion may continue until there is no iron left except what is contained in the residual dust. In addition, iron, in becoming oxidised, increases in size according to the quantity of oxygen taken up, and so it expels, having no adhesion to the metal, the film of paint, and the latter in time drops to the ground. Repetition and convenience have almost caused it to be regarded by some as an axiom that so long as iron is placed in salt-water mud, the metal will be free from any deleterious corrosive effect, because a film of rust will form which will protect the iron from further corrosion, and because there is no constant renewal of the water which reaches the iron. Even if fresh supplies of water were prevented from approaching the surface, which is hardly the case, although *temporarily* it may be a kind of shield to the solid metal, it will fail from the causes stated.

On leaving the rolls oxidation may be said to have commenced. This can, of course, be arrested if the metal be dipped in boiled oil or tar-asphaltum preparation; but if it has to be *afterwards* worked or manipulated, any protective films or portions of the preparation that have been imbibed by the pores may be disturbed, and then these form nuclei around which corrosion proceeds.

Why does corrosion occur? The oxidation of iron being complex, requiring conjoint action or influences, and not being a simple chemical combination of two elements, the question cannot be answered in a short sentence. It is advisable to remember that there are two kinds of chemical affinity; one in which substances are capable of combining, one of the substances being in any proportion whatever to the other; and the other in which substances only unite in certain definite proportions. Solutions of solid substances in fluids are instances of the former, iron-rust and water that of the latter.

"Neither bright iron nor steel will rust in pure water or in pure air. The presence of carbonic acid, or some similar agent, seems necessary, although the final product may be destitute of carbon. Even when oxygen, moisture, and carbonic acid are all present, rusting will not, it appears,* take place unless the moisture condenses on the surface of the metal. When rusting does take place under ordinary circumstances, the first stage appears to be the formation of ferrous carbonate. This carbonate is then dissolved in carbonic acid water to form ferrous bicarbonate, which latter is then decomposed in presence of air and moisture to form hydrated ferric oxide, magnetic oxide being formed as an intermediate product."

Four compounds of iron with oxygen are known.

	Iron.	Oxygen.	Colour.
Protoxide	1	1	Green hue, changing to red brown.
Sesquioxide (peroxide) ..	2	3	Blood red.
Black, or magnetic oxide†	3	4	
Ferric acid	1	3	Deep crimson.

---

* See ' Engineering,' February 9, 1894.
† The black oxide scales which form on wrought iron when heated to whiteness, as in the operation of forging, are chiefly composed of this oxide.

What is the cause of iron becoming corroded when exposed to moist air? The necessary moisture being present in the air brings into action the affinity of oxygen for iron or steel, the oxygen of the air combining with the surface of the metal and producing two kinds of oxide of iron, the ferrous oxide, which quickly becomes converted into the ferric oxide. Its durability, therefore, chiefly depends upon its power to resist combination with oxygen. Dry air at a high temperature is not encountered in the majority of engineering structures, and therefore need not be considered at length, although even dry air oxidises iron at a high temperature, as is proved by the familiar example of a red-hot poker being found to be rusty when it cools. It has been found in warming buildings by steam that if air is inside an iron vessel or tube in which steam is condensing, iron corrodes rapidly. If the air is excluded it does not. Then external corrosion has to be guarded against more than internal decay. Lead slowly absorbs oxygen and carbonic acid in moist air, and copper slowly oxidises in a moist atmosphere, and a green carbonate is formed on its surface. Experiments have shown that mercury, like many other metals, does not undergo oxidation either in dry air or in aqueous vapour, but only when aqueous vapour is present in the air, then it corrodes rapidly and oxide of mercury is formed over the surface. This, it is considered, points to an electrolytic explanation of the oxidation.

Although water is decomposed by contact with iron, it is chiefly the oxygen contained in water which causes corrosion, and it is accelerated as the temperature increases, and by absence of light. Some moisture is necessary, although its decaying power is dependent on other coincident circumstances, such as climate and character of the air, but moisture is an all important element in the production of corrosion. For instance, steam pipes coated with mineral wool or slag wool are kept free from rust so long as the wool is entirely dry, but where moisture is present it permeates the wool and reaches the pipe, and its surface becomes corroded.

Professor Egleston (U.S.A.) found that when moisture is at all constant in blast-furnace wool there will be a decomposition of the slag, and an attack on the iron by sulphuric acid set free, or by organic acids if the material comes from the drainage of the soil.

Oxidation which may be comparatively slow in a moist atmosphere is very greatly accelerated in air or water containing acids or other corrosive agents, consequently it is the climatic conditions which principally govern the rate of corrosion, and account for such variableness in the serviceable life of structures in which steel or iron are employed. The appearance of corrosion on the surface of iron is familiar, although it assumes various forms, as a skin of green-brown, brown, or blood-red, or deep crimson rust, nearly circular raised lumps having a centre of energy underneath which will be found depressions in the plate, or bunches or flakes of red-brown rust, which can be easily detached. That of steel is not so well known, thus the lilac tinge in steel is produced by partial oxidation, and corrosion has commenced. Why does the process of oxidation continue and ultimately destroy an iron or steel plate, and reduce it to a condition of practically being mere dust ? Without attempting to explain the chemistry of the matter, it may be stated that in iron the film of oxide of iron on the surface being converted into ferric oxide, it becomes an active corrosive influence, inasmuch as part of its oxygen is transmitted to the surface beneath to form ferrous oxide, and this process does not cease, but is repeated until the whole of the iron is oxidised, and then soon becomes a mass of mere flakes of fragile scale, which do not adhere, and ultimately become dust. Therefore, the importance of preventing any oxidation is indisputable, more particularly as reliable experiments have shown that the corrosion of plates of wrought and cast iron, and especially of steel, rapidly increases with time.

In the very earliest years of modern civil engineering, Smeaton wrote on corrosion, " I had observed that when iron once gets rust, so far as to form a scale, whatever coat of·

paint or varnish is put over this, the rust will go on progressively under the paint."

It may be regarded as an axiom that when rust has
commenced it will proceed. The hygroscopic character of
the hydrated ferric oxide causes an absorption of moisture
from the atmosphere, and so under ordinary circumstances
the chief necessary element is present, and it may be aided
by the rust, which being electrically connected with the iron,
will result in galvanic action.

Oxygen alone does not cause corrosion, as will be gathered
from the experiments of Dr. Grace Calvert, F.R.S., who, in
describing some he made with perfectly cleaned blades of
steel and iron exposed for four months to the action of
different gases, in order to determine whether the oxidation
of iron is due to the direct action of the oxygen of the
atmosphere, or to the decomposition of its aqueous vapour,
or whether the very small quantity of carbonic acid which
it contains determines or intensifies the oxidation of metallic
iron, found that the blades showed the following results :—

> Dry oxygen : no oxidation.
> Damp oxygen : in three experiments, one blade only was slightly
>   oxidised.
> Dry carbonic acid : slight appearance of a white precipitate upon
>   the iron, found to be carbonate of iron.
> Dry carbonic acid and oxygen : no oxidation.
> Damp   „   „   : oxidation very rapid.
> Dry and damp oxygen and ammonia : no oxidation.

These experiments tend to show that carbonic acid, and
not oxygen or aqueous vapour, is the agent which determines
the magnitude of the oxidation of iron in the atmosphere.
Therefore, carbonic acid in a damp atmosphere is a powerful
and active agent in causing corrosion, and it was further
shown by experiments that when iron was immersed in
water containing carbonic acid it rapidly oxidised ; that it
was not due to the fixation of the oxygen dissolved in the
water, the occurrence of hydrogen collected above the liquid
in the test-bottle sufficiently proved, but it was due to

oxygen liberated from the water by galvanic action. Caustic alkalies were also found to much retard corrosion.

Although oxygen cannot *alone* be considered as the active agent of corrosion, for a piece of clean iron or steel can be exposed in a dry atmosphere for a long time without appreciable corrosion, and this notwithstanding oxygen is present, its active corrosive power, i. e. the force of affinity it has for the metal, is great, but is governed by a variety of circumstances. However, the conditions that render it the active agent generally more or less prevail, especially the overpotent one of damp or moisture ; therefore in very dry localities, as Siberia, or in the arid plains and deserts of the world, corrosion is very much slower, even to being almost imperceptible, than in countries near the sea and subject to an annual heavy rainfall or comparatively damp atmosphere. If the atmosphere is excluded from a bottle of distilled water a smooth clean piece of iron can be immersed without deterioration, as also in salt-water from which the air is excluded, if 30 grains of quick or caustic lime to each ounce of water is added.

It is a characteristic of oxygen, i. e. acid generator, that it is exceedingly active and will unite in some way or other with any substance that can contain it, and may alone supply the acid requisite for corrosion, and is the supporter of ordinary combustion and the vital part of the atmosphere ; for, as they have been created, neither animal nor vegetable life could be sustained without oxygen being present in the atmosphere, nor could there be water or even atmospheric air, nor clay, flint, potash, lime, magnesia or soda. Various complications are involved in considering the mutual influence of animals and plants on each other and on climate. The food of plants may be said to be chiefly carbonic acid, water and ammonia, and there is, as it were, a constant interchange of food between the animal and vegetable creations, the former helping to supply carbonic acid to the plants, and the latter oxygen from the under surface of the leaves. In fact, so long as man exists on this earth in its present state, it is

c

decreed that oxygen and carbonic acid shall be present. As carbonic acid (CO_2), which is always in the atmosphere in varying quantity, in presence of moisture, is so active a corrosive agent, what are the chief causes of it? The breath of animals, the decomposition of vegetable and animal matter (hence the importance of no vegetable substance resting upon a painted surface), limestone, chalk and all calcareous stones in which it exists in a solid form, the presence of an acid setting it free. It is well known that carbonic acid is most likely to accumulate to a noxious extent from the fermentation and putrefaction of decaying vegetable and animal matters, and that by throwing fresh slaked lime into such places it prevents the accumulation of carbonic acid by absorbing it, and produces carbonate of lime, and that when the lime is slaked all its carbonic acid is driven off by the heat, and that, therefore, it is free to combine with more, and that heavy rain also prevents its accumulation by dissolving it. It also renders water in which it is dissolved slightly acrid, and it is a narcotic poison, and although gases diffuse themselves through each other, carbonic acid gas is heavier than air, and therefore is present in some places in fatal quantities near the ground and bottom of wells; and although a dog will be at once affected by it, his mouth not being sufficiently high above the ground to be beyond its noxious influence, a man may walk over the same place and suffer no inconvenience. Obviously, there is more carbonic acid present in crowded cities and manufacturing towns than in the open country. The quantity in the atmosphere is however very small, it being approximately, according to Dr. Angus Smith, in the open parts of London, 3 parts by volume in 10,000 parts of air. In the hills of Scotland 3·3 per 10,000, and in London streets 3·8 per 10,000. In the streets of Manchester during fog 6·8 per 10,000. In close buildings, it averages 16 per 10,000, and usually from 7 to 10 parts in houses.

There are other agents, whether caused by smoke, the steam of locomotives, or heat, that accelerate corrosion, and

among them sulphuric and sulphurous acid; others proceed
from water or the atmosphere, for neither rain nor dew are
pure water, as they contain acids and salts in solution,
chlorine, and ammonia, the latter being especially found
wherever organic decay is in progress, as in farm-yards,
stables, sewage, &c. Any substances which on decomposition
cause free acids to separate from them act as corrosives, and
if they are subject to steam pressure the deleterious effect is
much accelerated. A carbonaceous deposit on iron or steel
also tends to assist corrosion by acting as a nucleus to retain
moisture and acids, and also by condensing gases in its
pores, and by inducing galvanic action, carbon being electro-
negative to iron.

The influence of local circumstances upon the durability
of materials is demonstrated in the Houses of Parliament at
Westminster. Great care was taken in selecting the kind of
stone so that it might be durable, and a magnesian limestone
was chosen as it was known a very large number of buildings
had been erected with that material in the middle ages, and
that they were in an excellent state of preservation. No
doubt the stone would have been almost proof against
corrosive influences if the Houses of Parliament had been
built in the open country, or in a country district, but to the
atmosphere of large towns it had not been subject in the old
buildings, and therefore decay occurred, which, however, has
been arrested by treatment and a coating. In the case of
iron or steel, corrosive influences would also be much
increased, for it may be said the magnitude of the corrosion
of the same metal is chiefly dependent upon the position and
circumstances in which a structure is placed, and in great
measure any difference in chemical action is so caused, hence
the importance of a thorough consideration of the location
and peculiarities of any district in which a corrosible
engineering structure is to be erected.

Corrosion is affected by the quality of the coal used. In
London, the coal burnt is of much better quality than in
manufacturing towns; thus, in Manchester, it is said, the

c 2

coal has about 2 per cent. of sulphur in it, or about twice
that of the coal usually consumed in London, and therefore
the amount of sulphurous and sulphuric acid is greater, and
such corrosive influences are increased, and are particularly
powerful in manufacturing towns, whereas in the open
country it is hardly felt; and is an addition to the other
corrosive agents, such as carbonic acid, moisture, and the
usual deteriorating elements present almost everywhere.
Coal and coke of very good quality contain about 1 per cent.
of sulphur.

Iron or steel immersed in water are not subject to such
severe decaying influences as when they are placed in a damp
atmosphere, or where there is little evaporation. Dampness
is not mere moisture, or one approaching saturation, and for
corrosive purposes may be defined as being that state in
which moisture and air promote vegetable and other life
conducive to rapid decomposition, especially when assisted
by a warm temperature and absence of light; the result
being a continual combustion or fading away of the substance
by chemical action. All plant life, whether microscopic or
not, has a powerful decomposing influence. The chief aim is
to prevent the spores of such life being able to find a congenial soil or one upon which they can germinate, and also
to prevent the particular conditions under which they each
exist. Slime, or dust in a damp state, is frequently a film of
microscopic or other plant life.

The capillary action that occurs in materials has an
important influence on their durability, and within its limits
decay and corrosion will be considerable. A contributory
cause of corrosion is vibration, as it reduces the remaining
strength of any weak place, greatly tends to produce crystallisation, opens any crack in and shakes loose any paint not
closely and firmly adhering, and it is a general deteriorating
influence.

Vibration is not only experienced in structures exposed
to a rolling load, but also in manufactories, warehouses and
more or less every structure, and its probable amount and

effect should be considered. The fact that steamships which are subject to constant vibration from engines, shocks of waves, &c., corrode quickly does not support the idea that vibration lessens corrosion, although the inside of vessels are not exposed to drying airs, and saline particles which get into the holds do not evaporate, and so the plates are much more severely tried than any rails. Rails in sidings corrode much quicker than those under frequent traffic as in a main line, but this is probably caused by their not being subject to concussion, and the rolling, abrading and cleansing action of the tires on the heads, to the dirtier or unclean state of the rails, to the corrosive influence of the general débris and deposits of a goods siding, the comparatively undrained state of the ballast, its somewhat necessarily soiled condition, and, it may be, the less durability and strength of the rails used.

Consideration of some of the causes of corrosion and decay is useful in order to ascertain the reason of its occurrence. Combustion may be described as a chemical action in which the union of one body with another is attended with development of heat, and, under ordinary circumstances, with an evolution of light.

The experiment of the combustion of iron and steel wire in oxygen gas is well known, and is an example of this. Iron by some authorities is not considered to decompose water in the absence of air at ordinary temperatures without contact with some substance electro-negative to it. Professor Tyndall's experiments and labours showed that air is laden with myriads of germs and agents of decomposition ready to settle down and develop upon matter suitable to their growth. Fermentation has been described as a change in the elements of a body composed of carbon, oxygen and hydrogen without nitrogen, but putrefaction as a change effected in the elements of a body composed of carbon, oxygen, hydrogen and nitrogen. Dr. Brewer has briefly and well explained that " the carbon, oxygen, hydrogen and nitrogen of the original substance, being separated by decomposition, reunite in the following manner: (1) carbon and oxygen unite to form carbonic acid ;

(2) oxygen and hydrogen unite to form water; (3) hydrogen and nitrogen unite to form ammonia. When bodies containing sulphur and phosphorus putrefy, the sulphur and phosphorus unite with hydrogen, and form sulphuretted and phosphuretted hydrogen gases." Vapour is an elastic aeriform fluid, which may readily be converted into a liquid or solid merely by change of temperature. A gas is an elastic aeriform fluid, which cannot be made to change its state, except by the application of artificial pressure and intense cold.

The reason of some things being solid, gaseous, or liquid is that the particles are closest in the solid, furthest apart in the gaseous, and the others are liquid; heat changing a solid, like ice, first into a liquid, and then into a gas, because it causes the particles to become further apart, thus ice melts into a liquid, and additional heat changes it into steam. Corrosive action increases with the temperature, and at freezing point Fahrenheit it is very little. The effects of temperature are very considerable in chemical action, and a certain degree of heat or cold is necessary to produce the greatest activity. Experience in Sweden indicates that it is the constant changes of temperature from heat to cold, or cold to heat, that weakens the iron in rails. Mr. Sandberg * states that this explains why the breakages of rails generally take place in the autumn or spring, and not when the metal is for some days at a constant temperature. Repeated changes of temperature, even from 65° F. to 212° F., produce, it has been shown by Mr. H. Tomlinson, F.R.S., a considerable amount of internal friction between the constituent molecules of the metals, which tend to modify some of its physical properties.' Steel is more sensitive to this influence than wrought iron. Mr. Andrews, F.R.S., has suggested that this may probably be the cause of the greater number of breakages of steel axles in India and Canada, &c., where variations of temperature are extreme. In the very large number of experiments of Mr. Andrews, F.R.S., the metal

* See 'Engineering,' January 13, 1888.

was kept at 0° F. for cold tests, and 100° F. for warm, and
the appearance of fractures of the warm test was of a *fibrous*
character, the cold had a more fine-grained and crystalline
aspect. Statistics have clearly shown that railway axle
fractures are much more likely to occur during very cold
weather than any other. In Russia, about 50 per cent. more
steel axles then fail. Experience on the Grand Trunk Railway confirms this. However, increased rigidity of the
permanent way no doubt contributes materially to the
general results, and not merely the metallic change. Sudden
changes of temperature will assist any corrosion which has
occurred. In another way, alternations of mild and very
cold weather aid corrosion, for they will cause water of condensation to drip. In a similar manner dew is formed, it
being the vapour of the air condensed by coming in contact
with bodies colder than itself, and so is rain, for it is cold
condensing the vapour of the air when near the point of
saturation which causes water to fall.

The variations of temperature have been registered to be
not the same on both sides of a metallic bridge. This may
cause a condensation of vapour; for instance, at an arched
bridge at Liège it was found on one occasion, between 8 A.M.
and noon, that the range was about 29° F. on one side, and
16° F. on the other. The face or outside girders or ribs of an
arch being the most exposed to the sun, wind, rain, &c., are
also subject to greater variations of temperature. In some
experiments on the arched steel ribs of the Morand and
Lafayette bridges over the Rhone at Lyons, it was found the
maximum temperature in the metal directly exposed to the
sun occurred two hours earlier than the maximum shade
temperature of the air.

Experience has proved that very small quantities of
carbonic, sulphuric and sulphurous acids will cause rapid
corrosion of iron or steel. Where the better qualities of coal
are consumed, as in non-manufacturing districts, there is not
so much sulphur in the air, and consequently the amount of
sulphurous or sulphuric acid to be brought down by rain,

and so deposited upon any metallic surface, is reduced; however, sulphur set free from any substance in the presence of moisture will act corrosively and deleteriously on iron or steel. The smoke, vapour and gases escaping from ordinary locomotives have been shown by analysis to contain carbonic acid, which is considerably heavier than air; carbonic oxide, which is a little lighter than air; vapours and sulphurous or sulphuric acid in greater or less degree according to the quantity of sulphur in the coal burnt, and in addition chlorine and ammonia; the ordinary oxidising agents being also more or less present. Chlorine, which is heavier than air, is often employed as an oxidising agent, as, owing to its great affinity for hydrogen, it has the power of decomposing water, and setting free its oxygen, which, at the moment of its liberation, will combine much more readily with other elements than it will when perfectly isolated, and as the atmosphere of tunnels or underground ways is usually in a damp state, the conditions necessary for its most active attack are present, as with certain other elements which are especially energetic in their nascent condition. Chlorine has very powerful affinity for most metals, and moist chlorine is more energetic in its action than the dry gas. Thus in tunnels, subterranean ways, tubular bridges, and wide over-bridges on railways, all the influences required for rapid corrosion are present, for the atmosphere is then usually more or less surcharged with moisture from the aqueous vapour escaping from the locomotives, and condensation is rapid, and the tendency of the aqueous vapour to take up carbonic acid and sulphurous gases has also to be considered. Dr. Grace Calvert's experiments showed that carbonic acid and moisture produce rapid corrosion, and in the case of the structures mentioned must be added sulphuric acid and chlorine, two very energetic corrosive agents. Therefore it is specially necessary to constantly protect iron or steel when placed under such conditions, and also to consider whether it may not be better to employ cement brickwork or Portland cement concrete, but *not* masonry, unless the stone employed

will resist the acids, in preference to a metallic system of
construction, and any increase in the first cost may be justified
by the much longer life of cement brickwork or concrete, and
the very little attention required to maintain their surfaces
in good condition.

It may be well to remember that disinfectants, as they
are used in many structures, generally act by means of their
primary ingredient, whether it is chlorine, sulphur, or nitrogen, producing by their decomposition in presence of
moisture, hydrochloric, sulphuric, or nitric acid, all of which
corrode iron or steel.

A very careful examination by Mr. W. Thörner of the
corrosion of the permanent way in the tunnels of the Weilburg
and Nassau Railway * was made in 1887. Eight samples of
the rust, drainage water, mud and calcareous deposits in the
tunnels were taken from all the localities and carefully analysed. The results showed that except where calcareous infil-
trations occurred, the composition of the rust in the tunnels
was substantially similar from all localities. Chemical
analysis proved the essential constituent to be ferric oxide,
partly present as hydrate, with variable, but always small,
quantities of silica, alumina, lime and magnesia; carbonic
acid, when present, being in connection with the two latter
bases. The abnormal constituent was sulphuric acid, or
more correctly an oxidised sulphur compound, which occurs in
all proportions from $0 \cdot 3$ to $7 \cdot 9$ per cent., the actual state of
combination being doubtful, but most probably in the form of
basic ferric sulphate. Nearly all the samples showed a slightly
acid reaction, but only gave a very small amount of sulphuric
acid when exhausted with water. No rust contained iron as
sulphide. The origin of the sulphuric acid may be most
readily ascribed to the oxidation of sulphurous acid derived
from sulphur in the coal and dissolved in the condensed steam
from locomotives. This, trickling down the walls, is absorbed by the ballast and slowly oxidised by atmospheric air,

* 'Stahl und Eisen,' vol. lx., 1889.

producing sulphuric acid, which, penetrating the rails by capillarity, attacks them, forming ferrous and ferric sulphates. This, however, does not account entirely for the result, for rust obtained from ironwork exposed in a garden at Osnabrück gave 0·8 to 5·5 per cent., and these the sulphurous acid in the exhaust steam of the locomotives could hardly contribute.

A further series of experiments showed that nitrous acid and nitrites are formed by the direct action of iron upon air or water, and these substances are well known to act most energetically in the conversion of sulphurous into sulphuric acid. The existence of free sulphuric acid in the exhaust steam of a locomotive was determined experimentally on a heavy goods engine travelling at a speed of 30 miles per hour. The compositions of the incondensable chimney gases were found to be :—

	First Experiment.	Second Experiment.
Carbonic acid	5·4	6·1
Oxygen	13·3	13·0
Nitrogen	81·3	80·9
Carbonic oxide	—	—
Sulphurous acid	—	—
	100·0	100·0

Free sulphuric acid, ammonia and ferric oxide were found, but no nitrous oxide in the water absorption apparatus used for collecting and obtaining the composition of the gases. The amount of sulphuric acid evolved per hour by the engine was estimated, after deducting that required to saturate the small proportion of ammonia present, at very nearly 5 lbs., a quantity which under appropriate conditions is likely to act very destructively upon ironwork. The action is likely to be strongest in tunnels where the rock is not very wet, and the exhaust steam is without means of rapid escape, and especially in those where the rock is poor in carbonate of lime. Where the ground is very wet and provision is made for drainage, the soluble gases are taken up by the water, and removed in a very diluted form, before they have much chance of doing harm. Carbonate of lime acts beneficially by directly neutralising the sulphuric acid and converting it into gypsum. It was observed that the heaviest rusting took place between rail and sleeper when both were of iron; the acid water being introduced by capillary attraction, forms layers of rust which by continual accretion attains a thickness of 0·39 to 0·59 inch. Wooden sleepers, on the other hand, protect the iron on account of their low conductivity, which prevents the precipitation of the acid gases, if the surface of the sleepers is sufficiently covered by the ballast.

The methods suggested for the prevention of corrosion in tunnels after consideration of the results of these experiments are: (1) Covering the ironwork as far as possible with heavy or so-called carbonised tar, *not* ordinary gas tar, or asphalt applied to the metal in a properly clean condition and renewed at intervals. (2) The precipitation being greatest at the coldest points in the tunnel, i. e. on the ironwork, the latter should be covered with ballast, leaving only the rail head exposed, small limestone being the best ballast for this purpose. (3) Effective drainage of very wet tunnels. (4) In rock tunnels, when the rock is poor in carbonate of lime, the ground should be completely covered with small limestone fragments, or when that is not sufficient, the roof and walls should be lime-washed from time to time. Milk of lime sprinkling is not recommended. (5) The engines should make as little exhaust steam as possible in tunnels, and dispense with its use wherever it can be done.

These experiments, being conducted especially with the view to the prevention of rust under such conditions, are particularly valuable. The rock, &c., tunnels were in coarse-grained dolerite, Devonian limestone, and shales. In sound greenstone, containing no carbonate of lime, the rusting was the strongest. In the cretaceous marls containing 80 per cent. of carbonate of lime, corrosion was very slight. The Devonian schists were poor in lime. One sample of rust was from the Underground Railway, London.

The principal impurity in the Mont Cenis tunnel is carbonic acid, and there is an excess of aqueous vapour, and corrosive influences are considerable. The carbonic acid sinks to the lower part of the tunnel, and therefore the air the workmen breathe is affected, and not that of passengers in the carriages.

An analysis of rust taken from a bridge on the Pennsylvania Railroad, made by Mr. Kent, of the laboratory of the Stevens Institute, U.S.A., showed that in the water solution, by which method the tests were made, it was found that iron, ammonia, sulphuric acid and traces of sulphurous acid and chlorine were present, and carbonic acid in considerable quantity. These are readily formed by the fuel and steam of locomotives, and when they are dissolved in moisture, an acid and saline solution is formed having a powerful corrosive effect. Some experiments made by Mr. Kent indicated that "sulphurous acid is rapidly changed into sulphuric acid in the presence of iron and moisture, and that the iron is thereby rapidly corroded." The sulphurous acid escaping from locomotives must therefore be considered one of the most active agents of the corrosion of railway bridges.

The weight of vapour which will saturate a cubic metre (35·317 cubic feet) of air, at different temperatures, is as follows :—

	Centigrade.		Fahrenheit.	Grammes.		Lbs. avoirdupois.
At	− 10°	or	14°	2·302	or	·0051
,,	− 5°	,,	23°	3·406	,,	·0075
,,	− 0°	,,	32°	4·915	,,	·0108
,,	+ 5°	,,	41°	6·845	,,	·0151
,,	+ 10°	,,	50°	9·445	,,	·0208
,.	+ 15°	,,	59°	12·860	,,	·0283
,,	+ 20°	,,	68°	17·311	,,	·0381
,,	+ 25°	,,	77°	23·067	,,	·0508

To absorb the aqueous vapour without sensibly altering the hygrometric condition of the air, a difference in the temperature inside and outside the Mont Cenis tunnel of 15° C. in winter and 5° in summer is sufficient, and as this always exists, no additional quantity of air is required for its absorption.

The gases from the locomotives passing through the Arlberg tunnel, although great care is taken in the selection and combustion of the fuel, have, it is stated, caused it to be necessary to renew the whole of the ironwork after being ten years in use. The traffic is not considered to be great.

The atmosphere contains nitrogen, oxygen and numerous other gases, such as carbonic acid, nitrous, sulphurous, and sulphuretted hydrogen, ammoniacal and hydrochloric gases, varying very considerably in different localities, and in the same, according to the height above the ground. Some gases are especially to be found in certain localities, for instance, carburetted hydrogen in and around marshes, nitrous acid in districts subject to storms. Generally, town air contains more ammonia than country air, except in such special places as farm-yards, &c. ; and night air more than day air. The air of a sea-shore has generally a marked proportion of hydrochloric acid in it.

CHAPTER III.

THE CHARACTER AND QUALITY OF IRON AND STEEL, AND ITS RELATION TO CORROSION.

It is generally agreed that the magnitude and rate of corrosion much depends upon the quality and homogeneousness of the iron, for if films of oxide are present in the metal they come in contact with the iron, and then local action is set up and corrosion is uneven. It has been found in cable work that annealed iron of good quality does not deteriorate in sea water nearly so much as the inferior makes do, and that a less mass of the best iron is a better protection against corrosion than a much heavier one of inferior quality. It has also been found that hard cast iron with an even close grain resists corrosion to a considerable extent, but that soft foundry cast iron quickly corrodes, and that the whiter descriptions of cast iron corrode slower ; thus white cast iron of good quality resists corrosion better than grey cast iron, and the latter better than soft foundry iron, but grey cast iron is to be preferred for other reasons.

The Admiralty have for many years recognised that corrosion varies with the quality of the material, and have fixed certain chemical tests for steel or iron, which, if the metals satisfactorily sustain, are found to be safeguards against accelerated corrosion, and the metal then requires only ordinary care and the adoption of the usual means of preservation to prevent corrosion. In fact, want of uniformity in the composition of iron or steel resulting in laminations and imperfections in wrought iron, and heterogeneousness in steels, is now acknowledged to have a decided influence in promoting local corrosion. Also impurities in those metals within reasonable limits, that is, so as not to interfere with other necessary qualities, such as working and tenacity, are considered to affect their durability. It sometimes occurs that all the plates of a structure seem to be corroded equally ; it also happens that one plate may be more corroded than another, although apparently subject to precisely similar influences and conditions ; the only reason for such variableness in the magnitude of the corrosion is considered to be the want of homogeneousness of the metal, variation in its electric condition, the proportions of carbon, &c., being dissimilar, and one plate or bar having been manipulated, whether by hammering, bending, or in other ways, more than another, for the effect of bending iron or steel is to compress the particles or fibres on one side of the neutral axis, and to elongate them on the other.

Examination of the *bared* iron of a number of painted plates that had been submerged, has shown that the variation in the degree of corrosion was almost unaccountable if it did not arise from want of uniformity in the quality of the plate, for while some plates are equally affected, or almost free from corrosion, others will be very variably attacked ; local action, whether galvanic or not, having taken place to promote corrosion. In boilers, for instance, hardly any two similarly corrode, some rusting quicker than others. Constant care in the manufacture in order to maintain the *same* quality and purity of the metal is, therefore, one of the chief factors in preventing corrosion, and the greater the purity, homogeneousness, and freedom from foreign matters of the metal, other conditions being similar and in accordance with specified requirements relating to strength, the less will be the corrosion.

It is well to remember that what may be called commercially manufactured metals used in construction are not generally in an absolutely pure state. This arises not from intention on the part of the maker, but from other causes, such as the process of manufacture and contact with the vessels and the necessary plant used in its production.

A difference of surface caused by methods of production of the metal will lead to a variation in the rate and character of the corrosion, not only because of the alteration of the surface or skin, but also because of the difference of texture. The efficiency of the processes of manufacture from smelting to rolling, whereby hard and soft particles in the mass may be avoided, will affect the homogeneousness of the metal and the rate of corrosion, and complete fusion or melting are necessary to obtain it. The extreme quantity of metallic iron in commercial bar iron is about 99·8 per cent., but ordinary commercial iron contains about 98 to 98·3 per cent. of pure metal, the alloying elements being phosphorus, silicon, sulphur, carbon, manganese, &c. In ordinary *cast* iron, the pure metal would be about 93·5 per cent., and there would be about 2·5 per cent. of graphite. The various compounds do not corrode equally with pure iron, for a different material is affected, hence the composition of a metal will influence the rate of corrosion. Care in the selection of the most suitable mixture to obtain homogeneousness is required so that metal is not used that may have been compounded without much regard to uniformity of texture and character.

Analytical chemists of the highest authority are agreed that the presence of other metals in small quantities, such as iron, tin, zinc, arsenic, and the like, promotes the formation of insoluble scale on copper. In the Minutes of Proceedings of the U.S. Naval Institute, 1886, an account is given of the corrosion of the copper of the *Juniata*. Several plates, it was found, on the vessel being put in dry dock in October 1882, were so corroded as to be nearly or completely perforated, the immersed surface having become covered with a pale green, earthy-looking film, which blistered, many of the blisters split, and the coating flaked off to a great extent. It was considered that the copper beneath the spots of scale was softer than the surrounding copper, and made the difference required for setting up the corrosive action, and that the corrosion was principally due to the presence of spots of oxide of copper on the surface of the plates at the time they were put on, and that the action arose from unsound cakes of copper being taken for rolling, which contained cavities or air cells such as are occasionally produced, and that by rolling, the cavities are extended in the direction of the length of the bar, and when the bars are rolled into sheets, in the direction of the width of the sheet.

In granular iron, the degree of homogeneity, purity, and density appears to be considerable, and the smaller the crystals, the harder it is, and corrosion is likely to be reduced. In fibrous iron, the cellular and imperfectly amalgamated condition is considered to be caused by lowering the temperature to such an extent that imperfect welding results, as the bar is drawn out of the rolls in too cold a condition for the particles to unite; they consequently slide on each other, the result being fibrous iron, whereas it is declared granular iron requires a high temperature to produce it, and then the iron is homogeneous, or, in any case, much more so. It has been affirmed that iron should be granular, as it is the essential property of welding and amalgamation, and that all other textures are the result of defective and imperfect welding in the process of manufacture. However, other matters may influence the result, such as the composition, rapidity of rolling, and pressure.

In a fibrous metal, rust seems to eat its way into the laminations or sandwich of iron and dirt, as Dr. Siemens termed it, and forms a way for the entrance of corrosive influences. In a homogeneous metal, corrosion appears to be more even and regular over the surface, and therefore more of a granular nature, but even homogeneous metal is subject to pitting from corrosion; in fact, in any compound substance, corrosion will generally not be even and regular, and it can hardly be otherwise; however, the more a metal partakes of a homogeneous character so is corrosion likely to be uniform. Bad welding, unequal working, hardness, roughness, or polish of the surface will also cause a difference. In steel, uniformity can be approached more than in iron; however, the cinder in iron and the scale of steel are different. The former is a dielectric glassy substance and has no corrosive influence, whereas the scale produced in rolling steel has a very deteriorating effect, as it is a magnetic oxide negative to the steel, and therefore corrosion is rapid, as also if such scale is rolled into steel plates. Hard and crystalline irons are generally considered to be less oxidable than ductile or fibrous iron. Professor Kick, of Prague, after many experiments, which he only regards as indicating the probable general rule, not as proving it, with nitric, sulphuric, and hydrochloric acids and their combinations, with mordants composed of the salts of copper, &c., found that soft and fibrous iron of very good quality, also fine-grained iron, when attacked by the acid, is uniformly so, and with a limited elimination of the carbon, the surface retaining a dull lustre, a few incised specks and cinder-like holes being only observable. Coarse-grained iron and hot-short iron were found to be more powerfully attacked. In 10 minutes the surface became black, in half an hour, a black muddy deposit appeared, which could be washed off, and a number of small holes were distributed over the surface. This iron was not uniformly attacked, some portions being more deeply affected than others. Malleable iron or annealed iron rusted quicker than wrought iron, the action of the acid being very powerful and irregular. In Bessemer steel and cast steel, the action of the acid produces very fine fissures. In cast iron, the attacked surface presented a tolerably uniform grey colour.

Imperfect welding some consider is due to the defective mobility of the mass, the consequence of insufficiently high temperature to admit of the union of the particles. Welding in a non-oxidising atmosphere may be regarded as a step towards perfection. The electric welding of angle, T-irons and similar sections, pipes, is another; and, perhaps, the weakness of joints made in such sections of iron may be so lessened, and collecting places for corrosive influences be thus reduced; but there has hardly been time to determine the effect of electric welding on the corrosibility of the metal.

It is well known that there is a diminution of cohesive strength consequent upon welding. Some experiments with welded bar-iron have shown this to be about 20 per cent., and that resistance to sudden strain and to the action of accumulated work is, as a rule, even more impaired by welding than the resistance to constant pressure.

Material that is almost identically composed frequently has its powers of durability affected by its crystallisation. Quartz and flint, marble and chalk, are examples of this action. Any operation that deteriorates the coherence or density of a substance aids corrosion, because it not only affects the coherence of the mass, but separates the molecules and exposes them to the active agents of decay.

In making up a bloom by ordinary piling, if the piles are not of *uniform* quality the finished iron cannot be uniform, and therefore corrosion will be unequal, and while the edges may be irregularly corroded, the surface of a plate will be pitted, and bars may become furrowed; but even assuming the piles are of uniform quality, the surfaces of the bars which make the bloom, in being heated, have scale formed upon them, and although some of this skin or scale is removed by subsequent processes, such as by hammering and pressing, a portion remains in the iron after it is rolled and in its finished state, the dark streaks when iron is broken, notched, or manipulated in a lathe showing this. Apart from any moisture being confined upon the surface and the blistering of the paint, which are causes that may induce local corrosion distinct from any inherent properties of the paint, the unequal corrosion of wrought iron may be said to principally arise from electric action taking place between the streaks or veins, or portions or particles of scale incorporated with the metal in the processes of manufacture, which causes the mass to be one possessing varying electro-

D 2

chemical properties, consequently electric action can be generated. As every metal is electro-positive to its own oxide, and as the positive metal is more quickly acted upon than the negative metal, it follows that the former must be deteriorated, and that decomposition will not be uniform in consequence of impurities causing local action ; and further, there is nothing to prove that this internal action can be arrested or suppressed; in fact, reliable experiments support the belief that when electric action is set up, it will continue for a long time, and may not cease from various causes, and may even change in character ; it cannot be said, however, that so far as the question of films in wrought iron are concerned and their *continued* action on the mass, it has been determined whether the electric action ultimately ceases or not, although it becomes less.*

Iron in cooling does not crystallise in the same form, for its structure varies to some extent according to its composition, therefore corrosion will not be exactly the same. The microscope shows that crystals, streaks, and what may be called needles, spots, regular shaped figures and cavities can be discerned. Corrosion can, therefore, hardly be considered as likely to be uniform. It is known the segregation of substances held in solution requires time. If it were possible to solidify liquid metal in a moment, it would be more homogeneous than when it is gradually cooled.

In situations exposed to the atmosphere, water, rain, &c., and the corrosive influences which may be said to almost always prevail in engineering structures, dense, compact materials of uniform texture should be used, and those composed of elements, in any case, not readily soluble, and if practicable, insoluble in water, should be adopted.

Although iron and steel are apparently compact, it is known there are pores in the mass, and gases condensed in them have been extracted. Mr. Thörner at Osnabrück made some experiments with iron, and they showed that the

---

* *Vide* Chapter VIII., on " Galvanic Action and Corrosion."

porosity of volume per cent. of mass in ordinary cast iron was 1·41; in different kinds of Bessemer steel, 0·41 to 1·20; in a locomotive tire ingot, 0·57; in a locomotive tire ingot intended for making hard forgings, 0·97, the remainder being rail ingots.   One described as basic ingot iron, 1·95; another as basic steel, 1·22 and 2·17.   It is considered the experiments showed the porosity of iron and steel is subject to considerable variation in the different kinds of steel, that although strength and porosity are generally in direct relation to each other, the stronger being as a rule the denser, several notable exceptions have been observed; and that these latter are consequent upon the irregular distribution of the pores through the mass.   The pores were found to be of microscopic fineness, and must not be compared with visible blow-holes.*   These experiments confirm the statement that corrosive influences may penetrate a mass of iron or steel.

It seems to be agreed that iron and steel absorb at high temperatures gases which are partly disengaged when the temperature is lowered, and that the disengagement may be accompanied by an appreciable change in the chemical composition.   Some experiments of Mr. Parry (1872), and afterwards of Herr Muller, show that gases are enclosed in iron and steel; for instance, hydrogen, nitrogen, carbonic oxide, are present in them, and the gas by volumes in percentage of that of metal varies considerably, depending not only upon the nature of the metal but, as might be expected, upon the method of manufacture, for the cavities contain gas which is a poor conductor of heat, and so in cooling inequalities in texture and strength result and corrosive action follows.   In addition it is well to remember that moisture and air will penetrate the microscopic cavities in any metal.   Flaws in plates occur owing to defective rolling, &c.   Air, vapour, water, steam, &c., may penetrate into these from the ends of the plates or otherwise, and increase corrosion.   Castings are affected by cooling, contraction and

* See 'Stahl und Eisen,' vol. vi.

"drawing" of the masses, and various defects in moulding, such as the sand being too compact, and so not allowing the gases to escape from the metal through the sand, thus causing scabbing and honeycombing; also from it being too loose, a rough, uneven surface being the result. This is often the case in greensand moulding; the weight of a greensand casting is usually some 10 per cent. more than when dry sand moulds are used, consequent upon "swelling" all over owing to looseness of the sand and its capability of being compressed by the metal. If the whole mass were everywhere in absolute molecular contact it would be airproof, hence anything that is done to consolidate it and press it together and cause greater cohesion tends to reduce corrosion.

Unequal corrosion arises from a plate or any portion of a plate being more porous or less impervious to moisture or air than another; and without entering into the question whether it is possible or impossible for a liquid to pass into the interior of a piece of iron, it would seem that it is, for gases are known to be diffused through iron, and the sweating of the cylinders of hydraulic presses has been considered as a proof that such action is possible. It has been found when the natural skin or surface left in casting has been removed from cast iron, water under a pressure of about 3 to $3\frac{1}{4}$ tons per square inch will pass through the pores of the iron. It would permeate the upper surfaces and into the interior at much less pressure. In some substances, as wood, sponge, the absorption of water takes place with great force, and the power of capillary attraction has to be considered. The different processes of manufacture of iron or steel from the rough mass converted by hammering, pressing, or rolling, &c., show that cavities and interstices are present in iron and steel, and that attention to the uniform good quality of the metal and care taken in the manufacture can alone reduce them to a minimum. It may be accepted generally that the more porous and less carefully manufactured iron or steel will not only become corroded the quickest, but will be so more or less throughout the mass; whereas in a thoroughly homogeneous and very dense metal the corrosion would be superficial.

If any moisture in the surface metal to be coated was extracted by the paint, and the paint hermetically sealed it from air and moisture and gases of all kinds, corrosion, if any, could only proceed from gases or moisture in the pores of the metal, aided, as it might be, by galvanic action being set up consequent upon want of homogeneousness in the metal and the presence of impurities.

The texture of wrought iron will to some extent indicate the quality of the iron, for if it be good it will be fine, uniformly close-grained, and of a silvery grey colour; if common and inferior iron, it will have a coarse granular fracture and look something like cast iron, but the quality cannot be reliably ascertained from a mere inspection, not even if aided by the microscope, for mechanical and chemical tests are necessary. The colour of iron in its crude state is chiefly dependent upon the proportion of carbon in it, and therefore in appearance it is either of greyish tint, mottled grey, very pale grey merging into white, or nearly so. By many it is considered carbon is regarded as affording protection against corrosion. Wrought iron when bent is compressed on the inner and extended upon the outer side, the result being that any cavities on the inner are lessened while on the outer side they may be elongated, unless the reduction of area neutralises that action, which it will generally do to a small extent. The metal by bending is thus caused to be of unequal texture, and the more open portion will be liable to the most corrosion, if not carefully and uniformly protected. In some recent (1894) French experiments it was found that acids applied to irons or steels when the latter are strained above their elastic limit corrode the surface with greater rapidity along the lines of deformation while the metal is subjected to the action of a disturbing force, and that there are zones of deformation between which the metal is unaffected. This is probably caused by the disturbance and opening of the structure and grain of the metal and by its increased porosity.

There is no doubt that a specification which demands a metal of very good and homogeneous quality requires one which is much more able to resist corrosion than if the common kinds of iron were allowed to be used. However, although this is something gained, even the best qualities of iron or steel should be protected as much as possible against corrosive influences, and every care taken to ascertain that the paint or composition* does not contain anything that can corrode or otherwise injure the metal, such as acids, which oil, necessarily so much used in oil paints, frequently does. Mr. R. Mallet, in the early experiments made on the corrosion of different kinds of iron in pure and foul atmospheres, and in clear, foul, acidulated, and salt waters (see his original reports), found that when the metal was exposed to water holding air in combination, the surface corroded uniformly or in patches, either by rust or conversion of the original iron into plumbago, and that the extent of the corrosion depends upon the want of homogeneousness of the surface or metal, or its density or hardness, or in the combination of the carbon with the iron. Mr. Andrews' recent experiments have shown that it is important to guard against accidental galvanic action in structures, and that bars, plates, rivets, bolts should be as near as practicable of a similar material, temper, and composition, &c.

With regard to the method of manipulation and uniformity of the texture of iron and steel, hydraulic forgings are to be preferred to hammered forgings, for work done by the press is more accurate, destructive vibrations are not caused, and the force of the press is transmitted in its entirety to the ingot, whereas the useful effect of the hammer is not more than a comparatively small percentage of the theoretical effect, because of the elasticity of the blow. Press-formed ingots are usually finer towards the centre,

* *Vide* Part II. of this book, on 'The Prevention of Fouling and Corrosion in Submerged Structures and Ships.'

those produced by the steam-hammer are poorer at the centre. In hydraulic forgings of large size, the action of the press causes the pressure to be more equal without jarring, and it is brought upon the centre of the mass, whereas ordinary steam hammering causes it to be chiefly on the surface, and then not always equally so, and the centre of the mass is perhaps hardly affected. Compressing the liquid metal in the mould immediately after casting, and substituting an hydraulic press for the hammer in the subsequent forging of the metal, makes it of closer and more uniform texture, and therefore it is less liable to corrosion. A heated ingot may resist the blow of a hammer, but the hydraulic press continues to act until it has forced the particles into closer contact, and uniformly so to the eye, whereas the effect of a blow is but for a certain distance and it then ceases, the action of steam-hammering being superficial as compared with that of hydraulic power.

Similarly in the case of riveting, a light riveting hammer will not set up the centre of a rivet, or fill the rivet hole. A plying hammer may, and frequently does, but hydraulic riveters do; however, they cannot always be used. Uniform continued pressure is always to be preferred to any kind of concussion. As any voids in rivet holes will admit air and moisture, bad riveting is sure to result in accelerating corrosion, even if it does not cause it to appear. Taking into consideration that a girder may have been badly riveted, the corrosion and consequent reduction of strength will continue until at last its weakened condition compels its removal, which probably might have been averted if the riveting had been done by hydraulic power, and the metal carefully painted with a really preservative paint. It has been stated that corrosion may vary in strained and unstrained metal. As in the pin system of connections the strain is localised upon certain parts of a structure, whereas in the riveting method of connection it is distributed; there is an advantage, apart from others, in the rivet as compared with the pin system as being less liable to promote corrosive influences, always provided the riveting is properly done.

Rolled joists, and iron of similar shape, are of comparatively soft iron that will lengthen without cracking or tearing under the unequal strains it receives from the grooves in the rolls, and the quality is inferior to that used in ordinary bridge girders, and there is usually an appreciable quantity of phosphorus and sulphur in it. Such iron is generally more easily corroded than the better kinds of iron plates, bars, &c., used in bridge construction.

Some unexpected results have been noticed on copper where a thin strip of the surface has been compressed by marking with a tool or been polished, as when so treated it has not corroded in sea-water, although the untouched surface has. This effect is considered to be caused by the compressive action of the tool on its being drawn across the metal, which is thereby made more dense, and by the polished or smoother face preventing lodgment of water upon or against any minute inequalities of the surface and its penetrating the metal. It may be useful to here give a scale * of hardness of metals that was used in the laboratory of the Technical High School at Prague, composed of 17 metallic substances, arranged in ascending order from the softest to the hardest. The tests were made by drawing a cylindrical piece with a conical point six times along a polished surface of the metal to be tested : (1) pure soft lead ; (2) pure tin ; (3) pure hard lead ; (4) pure annealed copper ; (5) cast fine copper ; (6) soft bearing metal, copper 85, tin 10, zinc 5 ; (7) cast iron, annealed ; (8) fibrous wrought iron ; (9) fine-grained light grey cast iron ; (10) strengthened cast iron, melted with 10 per cent. of wrought turnings ; (11) soft ingot iron, with 0·15 per cent. carbon, will not harden ; (12) steel, with 0·45 per cent. carbon, not hardened ; (13) steel, with 0·96 per cent. carbon, not hardened ; (14) crucible cast steel, hardened and tempered, blue ; (15) crucible steel, hardened and tempered, violet to orange-yellow ; (16) crucible steel, hardened and tempered, straw-yellow ; (17) crucible steel, glass hard.

Vide 'Technische Blätter des Deutschen Polytechnischen Vereines in Böhmen,' vol. xiv.

CHAPTER IV.

CAST iron is a very heterogeneous compound, and care should be taken to ascertain that castings have not been made with the sole object of causing them to appear externally sound enough to pass examination. Iron of one kind, even if very good, is not found to be equal to metal made from judicious mixtures. Cast iron being crystalline, and wrought iron fibrous, corrosion is not similar. If cast iron is hard, not to brittleness or so as to decrease its working strength, of an even, close grain, with the carbon combined and not in the form of graphite, it resists corrosion to its greatest extent. If impure soft foundry iron, or the commonest mixture that has been called cast iron, but which is a mass composed of sand, cinders, scoria, or recrement, and soft, uneven, and open-grained metal having large crystals is used, it will corrode very quickly. On the other hand, close-grained, homogeneous grey iron will not readily corrode, nor will white iron of good quality and even texture. Cast iron when it is very rich in carbon is soft, like plumbago, and can easily be broken. Hence, when by corrosion it is becoming decomposed into such a state, its strength is gradually being destroyed.

The process of corrosion in cast iron is different in some respects to wrought iron, and in steel it cannot be said to be entirely the same as either. Cast iron slowly decomposes when it is placed in sea-water, the iron becoming dissolved or extracted from the mass, the remaining substance apparently occupying the same bulk as the original cast iron, there being no easily recognised reduction or increase of section as in wrought iron. The cast iron becomes of a graphitic nature, ultimately closely approaching the condition of plumbago or black lead, the material being subject to a molecular and chemical change more regular than wrought iron, and corrosion does not occur in similar streaks, grooves, or furrows, and is generally considered to be superficial and comparatively slow, but being one of decomposition by a chemical change of the mass, thus altering the character and composition of the metal, the depth of which, at present, cannot be ascertained in a structure without analysis, it should not be concluded because wrought-iron piles have become quickly corroded that all that is required is to substitute cast iron for the wrought iron, and that, because to ocular demonstration, the cast iron *appears* to be in as good condition as when placed in the work, no deterioration and decomposition have taken place. In fresh water, the corrosion of metal of good quality should be very slow and superficial, although by the ordinary laws of nature some decomposition will take place.

The process of corrosion being so different to that of wrought iron, or ordinary oxidation, which is apparent on the surface and ultimately reduces the size of the metal, it is necessary to be very cautious in determining the magnitude of the deterioration or decomposition, for a considerable loss of strength and cohesion may have taken place without any alteration of volume, or even appearance. This gradual deterioration of cast iron should demand regular inspection of the metal at intervals, and it is advisable that pieces should be submerged under the same conditions as the permanent structure so that they can be analysed and examined. In wrought iron, the reduction in strength may be said to be proportional to the measured corrosion, but in cast iron no deteriorating effects may be visible, and yet exist to an important extent; for instance, it was found by a high chemical authority in India, Dr. Lyon, on some pieces of cast iron being examined after $4\frac{1}{2}$ years' immersion in

Bassein creek, which water is pure undiluted sea water having a specific gravity of 1·028, and contains 3000 grains of solid matter per gallon, of which 1605 consist of chlorine in combination with sodium, magnesium, &c., that they had undergone a change to a depth of $\frac{1}{10}$ inch from the surface, resulting in a small solution of the iron forming this portion of the piles. In commenting on several such experiments, Dr. Lyon stated that in his opinion the action of sea water on cast iron consists in a gradual solution of the cast iron, leaving the carbon undissolved, and as the percentage of iron in cast iron varies within certain limits in different specimens, it follows that, unless the percentage of iron be reduced below these limits, such change must remain undetected unless the iron has been analysed before immersion. He considered cast iron so immersed will deteriorate in process of time, and it is a question whether it would not be advisable to guard the iron from corrosive action of the water by coating it with some protective material, or by rendering it electro-negative under the influence of some more oxidable metal, such as zinc.

Dr. Grace Calvert, F.R.S., made a series of experiments in acidulated water by immersing grey cast iron, 0·39 inch cubes, made of Staffordshire cold-blast iron. Specific gravity, 7·858. Weight of cubes, 237 grains. One cube only was placed in a corked bottle holding 31 cubic inches of much diluted sulphuric, hydrochloric, acetic, and phosphoric acids. The action of the acids on the iron was slow, but at the expiration of three months, although the appearance of the cubes of cast iron had not changed, some of the tubes, especially those immersed in acetic acid, had become soft to the extent that a knife-blade would penetrate the cube 0·11 to 0·16 of an inch. The solutions of the acids mentioned were replaced by a fresh supply in each bottle every month during two years, when changes had taken place in all the cubes, acetic acid having the most powerful decomposing effect, then hydrochloric and sulphuric acid; phosphoric acid not showing such a result. The acids had so acted upon the metal as to *change its nature, without alteration of its bulk or the appearance of its surface.* The weight of a cube was after two years' immersion but 54 grains, and the specific gravity only 2·751, instead of 7·858. It was found that the iron had either been dissolved or extracted from the mass by the action of the acid, and in its place was a carbonaceous compound of less specific weight, having very little cohesion, the bulk or dimensions of the cubes being the same as that of the original cast iron.

—	Analysis of cube of grey cast iron as described, *before immersion.*	Analysis of the same cube after two years' immersion in acetic acid, resulting in its becoming a carbonaceous substance.
Iron	95·413	79·960
Carbon	2·900	11·070
Nitrogen	0·790	2·590
Silicium	0·478	6·070
Sulphur	0·179	0·096
Phosphorus	0·132	0·059
Loss	0·108	0·155
	100·000	100·000

These experiments, although made under more severe conditions than those which occur in ordinary practice, are most instructive; for the acids mentioned are often present either in the air, ready to be brought down by rain, or are in the waters of rivers or the sea, and it may be said that either the acids, or the elements from which they may be formed by decomposition, are more or less present in all the waters and situations in which engineering structures have to be erected, therefore, all surface water should then be regarded as acidulated.

As under similar conditions conflicting statements have been made as to the corrosion of cast iron in sea water, perhaps the difference may have been due to disturbance of the skin on the cast iron or to the different quality of the metal, or to some local cause not noticed, and if observed, not thoroughly investigated. In one place cast iron is pointed out as being uninjured after about half a century's submersion or erection. On the other hand, some cast iron has undoubtedly had its nature changed, and become another substance, and been made quite soft and somewhat similar to plumbago. In any case it is a material to view with some doubt, because any corrosion cannot be seen as it can be on the surface of wrought iron or steel. It may be perfectly sound, and it may not be. It is not agreeable to be answerable for so unreliable a material; and it is one not to be too quickly adopted without full consideration of the consequences of failure.

The shrinkage strains in castings affect the homogeneousness of the metal, for the particles near and about the surface are able by the softness of the interior metal to get closer to each other in setting than those in the more central portion which become cooled more slowly; thus when the surface metal has become corroded, the interior iron will be affected much more quickly. Ribs and lugs cast on a cast-iron plate, as on a pile joint flange or bearing plate, have a tendency to curve the plate. If the ribs and lugs be thinner than the plate they will become cooled first, and then, in resisting the shrinkage of the plate, probably make it curve upwards. Where the greater mass of metal is, cooling and shrinkage will be last, therefore the thicknesses of a casting should be regular, and mouldings on columns are better omitted, or they can be attached to the columns as ornaments. The shrinkage of a casting takes place while it is changing from a red to a black heat. Homogeneousness in a casting is necessary to ensure uniform resistance to corrosion. The metal in castings cooling first near the surface has a close and equal grain, and is harder than the internal portion, therefore it is not homogeneous throughout the mass, for the outer and inner portions have not the same strength. In very thick castings the interior metal will approach a cellular or honeycomb form.   Mr. Barlow's experiments on beams showed that the strength of the skins of bars of British cast iron was about two and a half times greater than the interior metal. The importance of preserving from corrosion the outside skin of columns or castings is great, for if it becomes corroded or a change occurs in the character of the metal, such deterioration can only be regarded as very serious, as it is the strongest metal, and even if one side only apparently has so altered, the corrosion and deterioration will soon penetrate and envelop the outer skin.   In thinner castings, which, however, are liable to be more brittle, say those having a thickness not exceeding about ¾ inch, this inferior interior metal is not so evident.   Columns which have the most unequal thickness will have the greatest tendency to crack, and when a column is cracked, continued and increased pressure is sure to enlarge the fissures.   Corrosion will soon occur internally and in all the cracks.

In some experiments at the Royal Gun Factories, Woolwich, it was found that for columns vertical casting was the best system, as the scum of the iron is then at the top, and not on one side, as in horizontally cast pillars.   An extra length, equal to about the diameter of the column, can be cut off in the lathe.   If the casting is thick, the extra length should be a little more than equal to the diameter of the column.   The decided weight of practical opinion and experience is in favour of casting columns vertically under the pressure of a high column of metal in order to ensure their solidity.   The greater the solidity, the less the corrosion. Even when a column is cast on end, and a head of metal equal to the external width or diameter of the pillar is cast on, and afterwards cut off in the lathe, any projections or thicker portions of the pile may be hollow or porous.   To increase the soundness of castings a plan has been adopted by M. von Riet of fixing a separating chamber above the flask.   It is divided into three circular compartments.   The molten metal is poured into the largest of these in such a way as to give it a whirling motion, which causes the heavy metal to keep to the sides, the light scoria remaining in the centre. A communication to the second compartment is made at one side of the first, and here a further separation is effected in a similar manner, the purified metal escaping finally into the third compartment, in the middle of which is the pouring hole to the flask. On cooling, the first compartment is found to contain large lumps of cinder, in the second the iron left is of a spongy texture. The castings, however, are stated to be very sound and solid.

If a girder is cast on its side, the scum of the iron is on one side. On the other hand, when it is cast upright, that is with the bottom flange lowest, the scum is in the top flange, where it affects the strength the least, and can be provided against by a slight increase in the thickness of the top flange. Unequal corrosion of a girder can thus be prevented, whereas when the scale is on one side of the web and one end of the flanges, corrosion, if a girder be exposed, will be unequal, and galvanic action may be set up by nuclei being formed in the iron, and a much greater surface of the cast iron has not the same naturally protective silicious scale. In the process of casting, the molten metal fuses the sand upon the surface of the mould and causes a hard silicious skin to be on the iron, which, if not perforated, broken, or altered, but covered with a really anti-corrosive paint, is a decided protection, and acts as a shield against corrosive influences. If the skin be removed, broken, or injured, the natural protective coat is destroyed, and corrosion or change of condition of the iron may ensue, unless the paint is sufficient to prevent it and is periodically maintained in that condition.

What is technically called "scabbing" occurs sometimes in castings, i. e., the sand rises from the under surface of the mould during the time the molten metal is running into it. Metal mixed with sand cannot be as strong as solid metal, therefore there is local weakness and inequality. Great care should be taken to procure uniformity of the nature of the metal throughout a column, or any corrosion will be unequal. As the kind and quality of moulding sands vary, and different

E

systems of mixing and preparing the material are used, and as it is generally considered that the silicious scale is a preservative, as undoubtedly it is if *kept intact*, it is well to remember that in good sand the proportions of silica and alumina are usually about 94 and 5 per cent. respectively, with traces of magnesia and oxide of iron. Sand containing much metallic oxide, especially lime, is not desirable for moulding. It is evident that the resulting scale on the cast iron cannot always be of the same nature, nor have identical anti-corrosive powers, nor will any paint adhere or otherwise have similar resistance to decaying influences on different material. It is probable that the character of the scale on cast iron may have a not inconsiderable effect in determining the corrosion, and may partly be the cause of in one case cast iron apparently being perfectly sound after a structure has been erected many years, and in another quite soft or seriously deteriorated in strength. Inequalities in the scale will also occur unless the moulding sand be of *uniform* character, perfectly mixed and thoroughly incorporated. Such defects may be and have been caused by the sand in a mould being rammed too tightly in one place and too loosely in another, and by want of homogeneousness in the sand. They will produce heterogeneousness in the scale, and also in the surface of the iron, and perhaps in the iron. Moulding sand should be *equally* tough and porous. Mr. Hodgkinson recorded that some cast-iron girders he tested had a permanent set when loaded with from $\frac{1}{57}$ to $\frac{1}{80}$ of the breaking weight, and that any weight, however small, injures the elasticity of cast iron, consequently *any* corrosion or decomposition of the metal may have serious consequences.

Recently in the U.S.A., and, it is said, in some instances in this country, thin sheets of wrought iron have been inserted in the centre of the mould before casting, It was first done in the case of plates for cooking stoves, and is stated to render them unbreakable by fire. Large iron pipes have also been so cast. The wrought iron perforated sheets sometimes used are No. 27 wire guage, and it is said that with such a plate inserted as described, a ¼-inch cast-iron plate is equal in strength to that of a plate 1 inch in thickness. The contact of metals of different composition is not, however, to be recommended from the point of view of the prevention of corrosion, as galvanic action may be generated. Colonel Yolland's report, September 16, 1884, on the fall of a cast-iron railway bridge, declared that "it is perfectly well known that cast iron cannot be relied upon." For the girders of railway bridges it is now practically an obsolete material. Cast iron is being abandoned in the best designs even for arched bridges. Its use in engineering construction is likely to be confined to cylinders, columns and piles statically strained, inferior parts of structures, pipes, &c., and probably will ultimately hardly be adopted for anything having to sustain any considerable or varying strain, or in any position in which a sudden failure would be disastrous or even inconvenient. Defects in the composition of cast iron, air-holes, cold shorts, cracks, cinders, &c., and corrosion are not to be easily detected, and can only now be ascertained by mechanical tests and chemical analysis. Cast iron, unless very great care is exercised in its composition and in the method of casting, should be regarded as an insidious material that may have the appearance of solidity and durability, and be really a more or less honeycombed and decaying mass, no longer possessing reliable strength, compactness, or durability. The importance and necessity of its being well and carefully manufactured throughout the mass has been recognised by the greater skill and attention given to promote its homogeneousness and good quality in every respect, but steel and wrought iron cannot but be considered as more reliable materials both as to strength and known durability. It is not here intended to condemn the use of cast iron or any other material, for each has its use, but in most engineering structures great care is desirable in its manufacture, for local defects are more serious in their results than in wrought iron or steel, therefore motives of prudence demand cast iron should only be used for those parts of a structure in statical

compression, or when subject to very slight other strains, and
unless these conditions are complied with sudden failure may
result, the combined effect of overstrain and deterioration
from decomposition of the mass consequent upon corrosion.

With regard to the corrosion of steel, that of wrought
iron and iron generally being examined in Chapter II., steel
is a somewhat indefinite substance, for it is known structural
iron is steely in some degree. Dr. Siemens suggested that
the following definition might be found to answer all requirements, namely, that steel was a compound of iron with
any other substance which tended to give it superior strength,
and stated there are substances, compounds of iron, manganese, and silicon sold for steel, which corrode very rapidly.
It is the improper selection and use of the material which
causes any excessive corrosion, for under proper conditions
steel is quite as durable as iron. The structure of steel
depends upon its composition and the method of manufacture.
It has been shown by means of microscopical examination
that corrosion will be variable unless homogeneousness is
attained, for if it is not, there will probably be crystals,
insoluble and other substances, minute cavities caused by
contraction and other matter in such steel, and galvanic
action may be set up internally. Examination by the eye,
and especially with the aid of a microscope, shows that the
fluid-compression process as applied to steel ingots causes
homogeneousness in the structure of the metal, hence it is a
decided preservative against corrosive influences, but whether
the same uniformity and solidity of texture can be or is produced by other means cannot be referred to here ; however,
anything that conduces to such an end is to be regarded as
having a decided anti-corrosive effect.

Unless the composition and processes of manufacture of
either iron or steel are declared, its probable uniformity of
texture cannot be judged. Dr. Siemens has said that " mild
steel was really iron of the best character. It was produced,
not like puddled iron, in small quantities to be welded
together with the chance of enclosing foreign matter, and

producing irregular results; but it was produced in large masses, 10 or 12 tons of fluid substance, and there was every probability that such a material was uniform to the utmost degree. Practice had fully substantiated the fact that there was no material more uniform than very mild steel, but it is not specially adapted for girders, a somewhat harder steel being better adapted for the purpose, and the selection of the best kind of steel for a structure is a matter requiring consideration in each case." If the section becomes small, as in continuous girders, it would seem to be advisable to use a milder steel than in ordinary girders, so as to have metal of the very greatest reliability, and that will bear almost half its breaking strain without apparent permanent set, and that will bend rather than break. In cold climates experience has shown that it is advisable to use steel possessing very considerable tenacity and softness, whereas in mild or warm climates almost any kind of approved steel can be employed.

The effect of quality on corrosion is a somewhat disputed question. Some consider mild steel for railway bridges should be of great purity, have no foreign matter other than carbon and manganese, that the proportion of carbon and manganese should never exceed a certain limit, the metal have a silky and not a granular fracture, however the appearance of fracture may be altered by the nature and manner of the strain applied, and that apart from questions of greater suitability of the steel for bridge work, corrosion is then likely to be less active. But it is to be remembered Bessemer tire steel, for instance, should contain, to insure soundness consequent upon the process of manufacture, about 1·25 per cent. of manganese. To show the futility of comparing the corrosion of a piece of steel with that of wrought iron without due consideration of the composition of each metal and the manner in which it is manufactured, careful analysis of a number of experiments conducted by different skilled operators demonstrates that in common steel having the scale on it, and made without much care, corrosion commences from a number of nuclei and not equally, whereas in steel of the most approved composition and carefully manufactured so as to be homogeneous, and the scale removed, the corrosion is more uniform and rather less than that of wrought iron of a similar class of metal. If what may be termed bad or inferior steel is compared with good wrought iron, the latter will appear to be better able to resist corrosion. If bad steel and bad wrought iron, or good and good respectively were compared, the steel would most probably be found to be the better able to resist corrosion, one reason being the absence of defective welding, which is a not unimportant cause of corrosion in wrought iron. It has not yet been proved, although it has been stated that steel in the atmosphere of towns rusts quicker than iron, but effective painting would prevent it, and the design of a structure should, if possible, enable it to be readily accessible for that purpose.

It is well known mechanical tests in every respect similarly made on metal chemically identical and from the same bar of steel will give different results. This can hardly be attributed to the heterogeneous chemical composition of the material. It may be to the extreme sensitiveness of steel of great strength to physical change, hence the tendency to use principally the milder steels in engineering construction. In fact, it has been stated that without annealing, it is impossible to get even in one bar an absolutely homogeneous molecular structure, consequently corrosion may be greater at one place than another, and corrosive streaks and pitting may be promoted in conjunction with other causes. It seems to be proved that, owing to strain, injurious molecular change may be very local. The results of experiments appear to indicate that annealed steel is much more soluble in sea-water than tempered or hardened steel; and that when they are in contact, the softer metal is rather rapidly corroded.

Some recent experiments on the Netherland State Railways, extending for about five years, 5000 trains passing over them yearly, tended to show that mild steel rails lost through wear and rust about 25 per cent. more than hard, but that rails of great hardness were considered as more liable to fractures; the lighter rails, 60 lbs. per yard, are specified to have a tensile strength of 33 tons per square inch, and the heavy rails, 80 lbs. per yard, 36·8 tons per square inch. As yet it cannot be said whether soft or hard steel is the more liable to corrosion, but it would seem that soft or mild steel is rather more corrosible than hard, but the different composition of the metal must always influence the results. Usually, the hardness of steel depends chiefly upon the amount of carbon in it, and the finer the grain the harder will be the metal and the more carbon in it. Considerable inequalities exist among different specimens of steel and iron, and therefore no fixed rate of corrosion can be established. The power of the corrosive influences will also seldom be precisely the same.

The reliable and recent practical experiments of Mr. Thos. Andrews, F.R.S., M. Inst. C.E., of the Wortley Iron-works, on simple corrosion, i. e. the test plates *not* being subject to galvanic action, in sea water taken from Filey Bay, and changed monthly, showed that as regards soft steel plates, the lower the percentage of combined carbon the less the corrosion, the best scrap iron corroding during a period of 110 weeks much less than the steels for the same time. The wrought iron, which contained roughly about double the percentage of phosphorus and manganese, corroded more than the best scrap iron, but less in the total period than any of the steels, with the exception of soft Bessemer. An excess of manganese in any of the metals is liable to produce increased corrosion, because it tends to induce local action from its uneven distribution. Weighing the test plates proved the correctness of the conclusions, and the results of galvanic action confirmed them. Further, limited to the question of corrosion only, it may be surmised that a body possessing a fine crystalline texture like steel, and having a complex chemical composition, would be ultimately more liable to ordinary corrosive disintegration than a fibrous substance such as wrought iron, and that in the construction of iron or steel ships, or any structures submerged in sea-water, or any kind of marine work, it is always desirable, in order to reduce internal causes of disintegration, to endeavour to have the bars, plates, rivets, bolts, as near as practicable of a similar material, temper and composition.

A certain percentage of carbon makes steel hard, but steel having comparatively little carbon is not so. Hardened steel has a very much finer texture on fracture than before it underwent that process, when it might be coarse. The question of variation of strength in the same chemically composed bar is one that has not yet been satisfactorily solved, although many reasonable inferences have been deduced, and at present it seems to be in the direction of change of structure that the solution of the problem is to be sought. It has been shown in practice that hardness alone does not always in rails necessarily produce great power of resistance to wear, and it may be considered as proved that hard and soft metal in other respects of the same character will not be similarly affected by corrosion. Careful and continued microscopical examination has shown that the crystallisation of steel does not necessarily proceed equally, and that in cooling down it undergoes a molecular change. Any variation in the metal may aid corrosion either by the formation of nuclei for such influences or by the production of galvanic action. The Société Cockerill, of Seraing, arranged their steel in four classes. *Extra mild steel*, carbon, 0·05 to 0·20 per cent., used for boilers, ship and girder plates. *Mild steel*, carbon, 0·20 to 0·35 per cent., used for railway axles, tires, rails. *Hard steel*, carbon, 0·35 to 0·50 per cent., used for rails, special tires, pieces subject to friction. *Extra hard steel*, carbon, 0·50 to 0·65 per cent., used for delicate springs, files, saws and cutting tools.

As aluminium is being tried and used in iron and steel manufacture, it may be well to here state that experiments have shown it is affected by acid and alkaline liquids and also by sulphuric and nitric acids, and therefore not to be particularly desired where considerable corrosive influences exist, although on high authority it has been otherwise declared. In some recent experiments of the Alloys Research Committee by Professor Roberts-Austen, C.B., F.R.S., samples of alloys were kept for some months before being analysed. During this time those which contained from 40 to 60 per cent. of aluminium had spontaneously disintegrated, and had fallen to powder. The powder was not oxidised, but consisted of clean metallic grains, probably resulting from chemical changes which had gradually taken place in the solid alloys. Whether the iron and aluminium were in a state of solution or were chemically combined when molten, there can be little doubt that they are so combined in the metallic powder, as attempts to remelt this powder have proved unsuccessful, which points to the formation of an infusible compound. Solutions are considered to be dissociated chemical compounds. These two metals, iron and aluminium, may have been too hot to unite when in the molten state; but at a lower temperature long-continued proximity at last effected their chemical union. It will hardly be supposed that this is an isolated case in the study of alloys. M. Le Chatelier, in a recent paper read before the Académie des Sciences, Paris, stated that equal parts of aluminium and copper were fused together in a crucible. The resulting ingot was then placed for 24 hours in a solution of common salt and lead chloride with a view to dissolving out the uncombined aluminium. No apparent change could, at the end of this time, be observed in the ingot. It was accordingly removed from the saline bath, washed and dried. At the end of 12 hours the whole mass was found to be reduced to the state of powder from the spontaneous oxidation of the alloy. A similar ingot, not immersed in the solution sodium and lead chlorides, was unchanged at the end of a month. It is understood to have been already found advantageous to paint the bottoms of aluminium-built vessels. In any case, they can hardly be considered incorrodible.

The French Government have used, and are using,

aluminium for torpedo boats. At the Norfolk Navy Yard, U.S.A., two aluminium plates were immersed in sea water 45 days. One was pure aluminium, the other contained 6 per cent. of copper. The pure metal was then only slightly affected, the other was much roughened by corrosion. They were again immersed and left for three months. The pure metal had suffered little, and was only slightly rough on the surface, and had few barnacles upon it, but the other was covered with them, and was more corroded. It is reported that the three small aluminium boats used by Mr. Wellman in his 1894 Polar Expedition, soon after being brought back could be easily crumbled in the hand. It would appear sufficient time has not elapsed for any absolute decree to be made as to the resistance of aluminium to corrosion in sea water. The purity, homogeneousness, and method of manufacture no doubt affect the results. Some considerable research has resulted in it being here expressed that it would be prudent, in now using it in sea water, to consider it to be, so far as regards corrosion in constructional engineering work, in a more or less tentative condition, and to protect it similarly to iron or steel when submerged in sea water.

When cold steel or steel at a low temperature is punched, the metal surrounding the hole is deleteriously affected, but if the metal is afterwards annealed the injurious effect is considerably reduced. As what may be called damaged metal is more quickly corroded than that which is sound, it has a doubly pernicious influence. The density of steel and the molecular condition are affected by treatment. Any kind of mechanical work upon steel has the effect of altering its physical character. Rolling down at a high temperature increases its specific gravity by compression of the molecules into intimate contact. The effect of phosphorus on steel is to make the metal brittle, liable to red shortness, and difficult to roll. Phosphoric steels are known to be untrustworthy when subject to severe sudden strains. They are less ductile and, it would appear, less homogeneous than ordinary steels, and therefore more liable to corrosive influences.

CHAPTER V.

NOTES ON THE RELATIVE CORROSIBILITY OF STEEL, WROUGHT
IRON AND CAST IRON, AND SELECTION OF THE METAL FOR
A STRUCTURE.

In determining whether to use steel, wrought iron, or
cast iron in any metallic structure, its relative corrosibility
should be considered. The preceding chapters have referred
more especially to the processes of corrosion in steel, wrought
iron and cast iron, and should be read in deliberating upon
the relative corrosibility, which has to be considered under
the circumstances in which the material is placed, apart
from that of cost, weight, size, strength, reliability and
adaptability for the purpose required. It may be affirmed
that a material which possesses the necessary elements of
resistance is to be preferred to one requiring protective
works, but, on the other hand, the cost of a structure made
of a certain material may equal, if it does not exceed, the
expenditure required by the adoption of a material requiring
protective works. This more particularly applies to river
and marine structures, which should be able to withstand
blows from shipping, barges, trunks of trees and logs, ice and
débris aided by the force of flood water.

The corrosion of wrought iron has been referred to in
Chapter II., and although it rusts somewhat more quickly
than cast iron, it has the great advantage that the corrosion
can be readily measured and the degree of deterioration
ascertained; nevertheless, when cast iron is used a greater
thickness is generally required, and considered from the
mere point of view of exposed surface to thickness, it has an

advantage, as comparatively little of the metal may be decomposed if protected by paint; in fact, it has been preferred to wrought iron simply because of this, the idea being that by increasing the thickness of the cast iron the molecular charge would be reduced. This may or may not be so. It has not been proved, but may logically be deduced; however, it is well to remember that other matters require to be considered, for increased thickness is not necessarily a proportional addition of strength. In the engineering structures herein more particularly considered, it is not advisable to use thicker castings than about $2\frac{1}{2}$ inches, nor thinner than about $\frac{3}{4}$ to 1 inch, the latter more particularly on account of the scale, the former because of the difficulty in obtaining homogeneousness, equal cooling, and accord in the order and direction of crystallisation; for it has been found wherever the order of crystallisation is disturbed there will be weakness and unequal corrosion, hence the importance also of as few abrupt bends, sharp angles and variations of thickness as practicable.

Solid forged wrought-iron piles are better able to withstand sudden blows than cast iron, but are considerably dearer. For tropical climates, general opinion seems decidedly to be in favour of cast-iron piles for submerged or the alternately wet and dry portions of a structure, as it has been found they appear to be in good condition, whereas the wrought-iron bracing, bolts and nuts occasionally submerged soon become deteriorated; however, the surfaces of the latter are greater as compared with the sectional areas, and they are more exposed generally. For the piers of lofty viaducts it is open to question whether cast iron is so well adapted as wrought iron, for other reasons than the mere relative corrosibility of the metals, as wrought-iron columns may be of sufficient diameter to allow of internal inspection, but, if they were of cast iron it may not be so, and the severe lateral strains on such a pier require to be met by a very complete system of bracing. In considering the deterioration of the metal, it is always well to be sure galvanic action is not the chief cause of any accelerated corrosion, and that it does not proceed from relative simple corrosion, if that expression can be used in regard to such a complex matter, for the electro-positive metal will become oxidised more rapidly than the electro-negative metal. It is very important to search for any interfering conditions in endeavouring to ascertain the relative corrosion of steel, wrought iron, or cast iron in any structure. To draw general conclusions from an isolated case is unadvisable, and has destroyed the apparent value of experiments and examinations conducted without regard to interfering conditions; however, if any experiments are conducted without such interfering conditions they are entitled to be considered as reliable, and are applicable to metal placed under the same circumstances in structures, but if otherwise, they may not be so. Those fully competent to judge have affirmed that by a judicious selection by an expert of steel and iron test pieces according to their chemical com-position, so as to take every precaution that they will show what is desired to be demonstrated, the steels might be declared by experiment to resist corrosion better than wrought iron or *vice versâ*, almost at the will of those conducting the experiments.   ·

The following question has to be answered in determining, solely as regards corrosibility, whether wrought or cast iron should be used for any purpose. Is it better to rely upon a material, such as wrought iron, in which the amount of corrosion can be seen, or upon a material, such as cast iron, in which the degree of decomposition cannot be measured except by chemical analysis and mechanical tests, although experience appears to demonstrate that such deterioration of strength and decomposition of the mass are both slow and superficial? No concise answer can be given to such a question, the various circumstances and conditions require to be carefully considered in all their different aspects. In submerged work, when the metal has simply to act as a column sustaining an insistent weight, and is not subject to variations of load or blows, cast-iron hollow columns are to be preferred to wrought-iron solid columns, their interior being properly preserved from corrosion. Many cast-iron columns are still apparently unaffected although they were erected from forty to fifty years ago in sea, estuary and fresh water ; however, there is no absolute proof by chemical analysis, or the usual tests for strength, that a portion of the metal has not become decomposed, or that decomposition is not proceeding, and the strength reduced and gradually diminishing, and a prudent course is to allow for such decomposition by increased thickness, within certain limits, and additional to any thickness required to meet the strains.

Solid wrought-iron piles are not often adopted for the columns of sea or river piers, and expert opinion indicates that for such structures cast-iron piles are to be preferred ; in fact, instances have occurred in which it has been considered advisable to substitute cast-iron for wrought-iron piles erected in sea-water in consequence of the corrosion of the wrought-iron piles having been very great, amounting to as much as 2 inches in a column originally of 5 inches diameter, in from ten to fifteen years, the piles being exposed to the air every tide, and the coast open.* This, however, it is advisable to consider as an exceptional case, and it is unfortunate that full particulars are not to be obtained of the reason of this severe corrosion, such as analysis of the iron, sea-water, whether galvanic action had been caused by juxtaposition of other metals, whether the scale was or was not removed, and whether there were any exceptional circumstances tending to promote corrosion or not, such as the constant discharge of sewage into the sea near the piles.

The late Sir J. W. Bazalgette once frankly remarked he had built a *wrought*-iron aqueduct for conveying sewage, and found it was not a wise selection, and had since adopted cast iron for similar structures, because, at any rate, a thicker material was thus provided at a less cost. The question of durability in cast iron Mr. Mallet defined as resolving itself,

* See Chapter VI., Notes on Rapid Corrosion, with Examples.

as regarded cost, into one of so much increased thickness at the outset as amply to allow for it in reference to a predetermined period of endurance.

For columns that are to be in the earth or submerged, and that consequently cannot be seen after erection, because of the greater thickness and somewhat slower rate of apparent corrosion, the statement seems to be justified that cast iron is a better material to use than wrought iron or steel; but if proper care can be taken in its preservation, wrought iron or steel is a more reliable metal, therefore the chief point for consideration is whether the wrought iron or steel can be made almost permanently secure by care in maintenance. The best material may become dangerous when neglected and allowed to waste away; but in works which permit of inspection and are carefully protected, wrought iron or steel are generally much to be preferred to cast iron, owing to their reliability and the extent of the corrosion being visible, and this even if the corrosion be somewhat quicker in action; bearing in mind that decomposition in cast iron may not be merely superficial or apparent, or capable of being measured until examination, analyses and test are made of a piece or pieces taken from the structure to prove that corrosion is merely superficial.

The durability of wrought iron or steel, so far as regards freedom from corrosion, may be said to be almost governed by the care taken in its preservation. Why is this the case more than with cast iron? Apart from relative corrosibility, the form in which the material is used is the governing condition. In cast iron, not only are the exposed surface areas very much less than is the case with a mass of plates, bars, angle, T, and other small sections of iron, rivet, and rivet-heads, but the silicious skin received in the sand mould can in great measure, by care, be preserved throughout, and the surface is also unbroken by rivet holes, projections and joints, around or in which moisture can collect and cause comparatively hidden oxidation of the surfaces or edges of the plates and other parts of a structure; therefore, although in the *mass* of metal there is not any change other than can be seen, there may be rapid corrosion between the plates, or underneath the angle T, channel irons, bars, cover-plates, and around the rivet-heads. Briefly, the relative exposed surface to the sectional area is much greater in a wrought-iron than in a cast-iron column or girder, consequently corrosion is increased, *unless the metal is protected.*

With regard to the relative corrosibility of wrought iron and steel, has experience shown that steel properly protected corrodes quicker than wrought iron under similar conditions and circumstances? So far as regards the safe employment of steel, it may be considered no fear need exist in using it. Occasionally a few alarming reports are disseminated to the detriment of steel, but upon investigation it is generally found either the circumstances were such that wrought iron would have been similarly affected, or the experiments were made under interfering conditions, such as galvanic action, which unfortunately render any conclusions that might be drawn from their results either right by good fortune, or misleading from the manner in which the tests were conducted. The original bloom or scale on test pieces has been proved to vitiate experiments. Several years ago the steel plates of a small vessel of very shallow draught were found to become rapidly pitted in the waters of the Irrawaddy, and a kind of frightful example was produced to show the inability of steel to resist corrosion; but it was shown that all *iron* vessels rapidly corroded in *that* river owing to peculiarities in the water.

Admiralty experiments demonstrate that there is practically no difference in the rate of corrosion between iron and the mild steels used by the authorities, provided, in the case of steel plates, they are divested of oxide on the surface. When under the same conditions as to corrosive influences and painting, if the scale was thoroughly removed, iron and steel corroded at about the same average rate, but steel more uniformly than iron. The scale being much harder to remove by chipping or scraping, after many trials, the immersion of plates and bars in dilute acid baths similar to those used in zincing iron proved to be the best method of clearing away the scale. It may be said all experts agree whether steel is used for plates, even when they are as thin as $\frac{3}{16}$ inch, boilers, or general shipbuilding or other purposes, the corrosion of the two metals is considered to be not very different, and any advantage is on the side of steel. A few qualify this declaration by stating that where steel is entirely immersed in water no more rapid deterioration occurs than in iron, but that in ships, between the light and load line, or wind and water, steel requires a little closer attention to keep the plates coated with anti-corrosive composition, in order to prevent its rather more rapid corrosion. In the U.S.A. some engineers consider steel to be much more durable, and less liable to corrode than wrought or cast iron. In this country perhaps not so decided an opinion would be expressed, although it would be in the same direction.

Some experiments made in France have shown that steel produced from pure iron, having hardly any manganese, carbon, phosphorus, or silicon, was less liable to corrosion than ordinary wrought or cast iron in the proportion of 0·4 to 1·4, being much in favour of a pure metal, such as very mild steel. Mr. Mangold, chemist to the Cologne-Müsen Mining Company, recently made * some experiments to ascertain the comparative corrosibility of iron and steel sheets, 1 millimètre = 0·039 inch in thickness, with the following constituents :—

	Iron. Per cent.	Steel. Per cent.
Carbon	0·16	0·06
Manganese	0·24	0·25
Silicon	0·72	0·00

When pickled in strong aqua regia (3 parts hydrochloric acid to 1 of nitric acid, specific gravity, 1·4) the steel was energetically and uniformly attacked. With the iron the action began slowly, becoming more rapid with the duration

* 'Stahl und Eisen,' vol. xv.

of the experiment, the surface produced was rough, showing signs of irregular welding. When immersed in spring water, the loss by two days' exposure was 0·04 per cent. in the iron, and 0·078 per cent. in the steel. The loss by ten days' exposure, the water being changed at intervals, was 0·09 per cent. in the iron, and 0·240 per cent. in the steel. When heated in a muffle at a bright red heat without access of air, the loss in 4½ hours was 18·32 per cent. in the iron, and 32·20 per cent. in the steel; and the loss in two hours more, 14·62 and 28·07 per cent. respectively. When exposed to an oxidising flame for three days both kinds were completely destroyed. When treated with a 1 per cent. salt solution, the loss in 24 hours in the iron was 0·037 per cent., in the steel, 0·128 per cent. The loss in three days was 0·085 and 0·220 per cent. respectively. When exposed to carbonic acid, the waste gases of a spathic ore-calcining kiln, and air, the loss in two days was 0·84 per cent. in the iron, and 0·94 per cent. in the steel, at four days, 1·60 and 1·40 respectively; and at six days, 1·95 and 1·90. When exposed to air and water vapour, the loss in twelve days was 0·29 per cent. in the iron, and 0·68 per cent. in the steel. When galvanised, wrought-iron sheets are more durable than those of mild steel. The latter can be more cheaply produced, and from the more uniform character of the metal have a better surface, but the zinc coating is thinner, and when it becomes unsound the plate is more rapidly corroded than is the case with sheet iron.

Provided the ordinary precautions are taken, similar to those adopted with wrought iron, experience shows that in considering whether steel or iron shall be used in any structure, their corrosibility may be taken as practically the same, but much depends upon the quality and texture of the metals, and therefore the strength, uniformity, and other qualities of steel are unbalanced advantages in its favour; however, the value of a protection from corrosion is greater in the case of steel than iron, because of the less section of metal required.

CHAPTER VI.

NOTES ON RAPID CORROSION, WITH EXAMPLES.

WITH regard to the conditions that cause rapid corrosion, iron in a very finely divided state, when exposed to the atmosphere, may oxidise so quickly as to become ignited. It is recorded that a cast-iron shell, which it was known had laid under water for about 200 years, when brought to land was found to be honeycombed, and some of the metal was in so fine a state of division that it gradually steamed, and became red hot; hence there may be danger in rapid corrosion under such circumstances, for decay may be said to be a process of more or less continual combustion. This is an instance showing that corrosive agents are especially active if they have not previously come in contact with a substance for which they have an affinity.

The smoke and vapour issuing from locomotives has a marked corrosive effect on girders under which trains pass, therefore, all hollows and depressions in which they can accumulate and hover about it is well to avoid, and to screen with a plate such portions of any girders. In sheltered cornices or nooks, or in ornamental work, or where projections occur, or in any places which will collect soot and dust and retain moisture, corrosive influences will be very active, and considerably more so than if the surfaces were smooth and even. Exposure to acid or alkaline vapour or liquid, to damp, occasional submersion in sea or fresh water, sewage, or the waste liquids from chemical works, considerable and varying heat and cold, constant or alternately occurring heavy rains with drying winds, hot sun, frost or snow, produces rapid corrosion, and also severely tries any paint used as a protective coating. Heating apparatus in buildings which causes damp places in the walls, or a state of humidity probably ending in vapour, has to be regarded as an active corrosive influence ; and it is advisable to consider it when girders, bearers, columns, &c., are subject to such heat, perhaps for a few hours during the day, and to the ordinary temperature of the internal air at other times.

Corrosion in fish-plates and metallic sleepers is generally most where the fish-bolts and fastenings occur. Inspection seems to indicate that the corrosion was much accelerated by loose bolts and fastenings, and it is difficult to determine whether the decay is more due to mechanical action and friction, or to simple corrosion, but there is no doubt such mechanical action greatly accelerates corrosion, even if it does not induce it. Mr. W. H. Cole, Assoc. M. Inst. C.E., states* that on the Sindh Sagar district of the North-Western Railway of India nearly the whole line, some 300 miles, was laid in 1886 with steel sleepers in sand, with a stone or brick ballast topping, the soil of sand and clay being impregnated with saline matter, but the air is generally very dry. The portion of the line that runs westward between the Salt range and the right bank of the Jhelum River towards the Indus is, however, for months exposed to inundation and is saturated by drainage from the hills. Sleepers weighing 148 lbs. each in 1886, averaged only 87 lbs. in 1890, or a loss of 61 lbs. in four years, or 41 per cent. Wooden sleepers were then substituted for steel sleepers. Guided by experiments with steel sleepers buried in sand, on the East Coast Railway, the authorities decided not to use them within ten miles of the sea. It has, therefore, been considered that steel sleepers should not be adopted in brackish soil, especially if it be moist.

A steel sleeper, being so thin, soon fails when attacked by rust. A cast-iron pot sleeper, being thick and apparently less liable to rust, although decomposition of the metal may

* 'Minutes of Proceedings,' Inst. C.E., vol. cxvii.

be proceeding, will last longer in brackish soil. It may be mentioned incidentally, as it may cause unequal corrosion, that trough sleepers, except when the permanent way has become consolidated in clean gravel or sand ballast, are difficult to pack on any embankment of soft or yielding earth, and their use is particularly undesirable at or near any points and crossings or junctions. While steel sleepers become soon rusted in some situations, it should not be forgotten that the white ant will quickly destroy wooden sleepers in certain districts.

From the effects noticed, some sewage water has been considered as corroding iron almost as fast as diluted vitriol. Cast-iron sewer penstocks, and the valve gearing, have been found in a few years to be rusted away through the chemical action of the sewer gas. A serious corrosive condition is always produced when raw sewage, or any sewage, or water made impure by human use, ammoniacal liquor, or chemical works, is discharged into river or other waters. Land frequently irrigated is in a condition to produce active corrosive influences, for it may be

> " Sated with exhalations rank and fell,
> The spoil of dunghills, and the putrid thaw
> Of nature."

It is generally soils that are the most permeable, and easily dried or warmed, that afford examples of successful irrigation, hence they are caused to be decidedly corrosive in their action.

The rainfall in an open district, as compared with a forest country, is much less, varying in different localities; added to which it has been found that evaporation under trees is only about one-third of that in the open ground, and that a covering of leaves will still further increase the difference. Metal-work in such localities will be subject to severe corrosive influences, in addition to any chemical action the rain or moisture may possess by extraction or impregnation with the juices of the trees and leaves; such water generally having a deleterious action on paint. Metallic structures erected upon

"Swampy fens,
Where putrefaction into life ferments,
And breathes destructive myriads ; or in woods,
Impenetrable shades, recesses foul,
In vapours rank and blue corruption wrapt,"

and upon marshes or bogs, will be liable to rapid rusting. It is therefore well to consider whether it is advisable to use some other and more suitable material than steel or iron.

Streams flowing from forest or wooded lands, or peaty soils, or any land having much decomposed matter upon or in it, generally have considerable sediment in them, and promote corrosion. By observing whether the water causes vigorous vegetable growth, and abounds with life, fish, or mollusca, indication is obtained that it will be decidedly corrosive. Salt-water marshes, lagoons, and estuaries are also to be classed as actively corrosive. It is well to here remember that carbonic acid is thrown off in great abundance by decaying plants in jungles, and the wind not being able to reach them, it therefore settles and destroys animal life, as in the Valley of Death in Java, and in some of the jungles of Hindustan, &c. Grass, wood, and the leaves of plants, radiate heat very freely ; rough woolly leaves radiate heat much more freely than the hard smooth polished leaves. By radiating heat freely, they condense any vapour which touches them into dew, and so a damp corrosive atmosphere may be formed in their immediate proximity.

The salts in sea-water are very hygroscopic, i. e. readily absorb and retain moisture, and have an energetic action on metals. Dust impregnated with sea-water may be blown about and appear to be quite dry, but when the air is charged with moisture it will become damp, and act as an active corrosive agent. Its effect, however, on macadamised roads is to bind and solidify them, and on stone pitching it tends to make them slippery. The employment of salt to melt snow on roadways, &c., will cause the resulting water to be actively corrosive; and if it reaches iron pipes it will have a deleterious effect upon them.

Cases have occurred in the same water, and under apparently the same conditions, in which cast and wrought iron have very quickly corroded, there being some peculiar influence, circumstance, or character of the metal or water which caused such a result. A perusal of this book may indicate some causes of rapid or exceptional corrosion, or enable them to be discovered.

It may be said all materials generally used in engineering works decay more rapidly when exposed to tidal action, or from a few feet above the ground, than they do when constantly submerged in ordinary sea or fresh water, or buried in earth in which foundations are usually placed. Anything which produces chemical action aids corrosion, as do fresh supplies of water or moisture, or the mixing of well-aerated fresh with salt water. In the latter case, corrosion in either cast or wrought iron or steel is very rapid. Mr. Beardmore mentioned an instance of this in the case of a sea-lock, in which soft water was "locked" down direct into very salt sea-water. When the lock at the head of a ship canal containing very soft water was pumped dry for repairs, after thirty-five years' existence, all the cast and wrought iron portions of the wooden gates, &c., were found to be equally destroyed, even the spikes of the platform planking had perished, though the timber was perfectly sound. Cast or wrought iron similarly exposed, and for the same length of time, in the Thames, he said, would be scarcely affected; and he attributed the intense action of the water upon the iron to the mixture of fresh and salt water. This corrosive action would be intensified as the saltness of the water increased.

Metal placed in a tank in which water is not changed will not corrode so quickly as if the water was constantly being emptied and renewed. The constant dripping of water on iron has also an active corrosive effect. Mr. Andrews's, F.R.S.,[*]

* ' Minutes of Proceedings,' Inst. C.E., vol. cxviii.

experiments have recently caused him to observe that, "Corrosion of metals is liable to occur in tidal streams, or under circumstances where the different parts of metallic structures, vessels, &c., may be exposed to the action of waters of dissimilar salinity. In tidal streams this state is brought about by the gradual rise and inward flow of salt water, and the outward flow of the fresh water. Hence, the upper and the lower portions of a metal structure or vessel, although composed throughout of the same metal, are exposed to electrolytic disintegration from the galvanic action set up by solutions of different salinity on the metal. Moreover, there are indications that magnetic influence tends to increase the corrosion of steel."

Mr. Edwin Clark stated that chemical action was invariably accompanied by electric action. It was, therefore, probable that differences in the electrical conditions, or rather the different electrical capacities of large and small masses, might explain any difference in their durability. He tested it by two pieces of plate 4 inches square and $\frac{1}{4}$ inch in thickness. One was attached to one of the tubes of the Britannia bridge by rivets, and was in direct metallic communication and formed part and parcel of the tube, but the other plate was separated from the tube by a thick glass insulating rod. The difference in the oxidation of the two plates was evident, as, in the course of eighteen months, the plate in communication with the tube was scarcely affected by rust, whereas the plate insulated by the glass arm, placed between it and the plate, was decaying rapidly. It was evident the more rapid decay of the small insulated plate was due to this insulation.

Iron or steel in tubular railway bridges, and in most railway bridges in a lesser degree, depending upon circumstances, girders in tunnels or underground railways, and similar structures exposed to the smoke, steam, and heated gases of locomotives, are especially liable to become rapidly corroded unless frequently tended under skilled direction. Although the maximum quantity of carbonic acid found in the

Metropolitan Railway was under 13 parts, or not so much as in close buildings, in railway tunnels the extent of acceleration of the oxidising influences is not to be measured by any excess of the quantity of carbonic acid only, as other agents of corrosion are present, such as aqueous vapour, sulphurous acid (SO_2), which is produced whenever sulphur is burned in the air or in oxygen, and is heavier than air, and experiments have shown that it quickly changes into sulphuric acid (SO_3) in the presence of moisture and iron, and that it is a powerful corrosive agent, and one of the worst met with in railway tunnels or bridges; for the ordinary trade sulphuric acid, or "oil of vitriol," which consists of one atom of the anhydrous acid combined with one atom of water, has a strong affinity for water, dissolves almost all the metals, and is the most powerful acid known, being, perhaps, the most important of all chemical products, and very largely used for manufacturing purposes.

It is open to much question whether the system of wrought iron or steel girders, with brick arches between them, is a preferable one to adopt in the construction of underground railways, because of the actively corrosive condition of such subterranean ways when worked by locomotives burning coal. Smoke and vapour will cling to and hover about the roof of a tunnel and the bottom flange of the girders, and any tie-rods are much exposed to corrosive influences. Sir B. Baker, in his paper "The Metropolitan and Metropolitan District Railways," * remarked, " In places wrought-iron girders were used, but though more trustworthy than cast iron whilst new, they are exposed to greater risk from hidden oxidation. Experience has shown the trouble and cost of maintaining ironwork exposed to the atmosphere of an underground railway to be such as would justify a considerable increase in the first cost, by substituting brickwork and deep cuttings for ironwork and shallower construction. Where the depth was sufficient for an arch, brick covered ways were adopted in preference to iron-girder

* ' Minutes of Proceedings,' Inst. C.E., vol. lxxxi.

constructions, on account of the smaller cost and increased durability." Sir John Fowler confirmed Sir B. Baker's views, and further said, "That with covered ways, and, indeed, all structures in London, it was wise to use brickwork when possible, and not ironwork, and even to incur some additional expense by so doing. Oxidation with iron and steel structures was always serious enough in London, but when hidden it might be dangerous." It will be noticed that no opinion is expressed, or experience recorded as to the relative corrosion of cast and wrought iron, but only that wrought iron is "exposed to greater risk from hidden oxidation."

It may be written because of the severe corrosive influences, general experience with ironwork in underground railways, in which ordinary and not electrical locomotives work, and in tunnels and similar subterranean ways, indicates that it is better to use brickwork in Portland cement, as being comparatively unaffected by the decomposing agencies which so seriously affect metallic structures. Wherever the atmosphere is of an exceptionally corrosive character, as is the case in some of the manufacturing districts, brickwork, and *certain* kinds of stone, not all, by any means, are to be preferred to iron or steel. As the air of London is not nearly so corrosive as some manufacturing cities and towns, such as Manchester, a stronger reason would prevail in any such localities for adopting some other more durable system of construction than that embracing iron or steel girders.

Corrosion will always be considerable in tunnels until smoke, vapour, steam, and gases, that vitiate the atmosphere or cause it to be actively corrosive, are prevented from entering. Some will always be present, even if sufficient power can be obtained by other approved means than steam. Among the chief causes of the disturbances of the air in tunnels, are the different atmospheric pressures, the temperature, and the winds. With very few exceptions, in long tunnels the atmosphere is saturated with moisture. The aqueous vapour from the locomotives surcharges the atmosphere and condensation is rapid. The aqueous vapour has a tendency to take up a considerable portion of carbonic acid and sulphurous gas, and bring it down when the vapour assumes the form of water. In tunnels the corrosion of rails generally increases as the ventilation becomes less and the length of the tunnel greater. The loss of weight by corrosion some consider as twice as much as that which corresponds to the wear of the upper surface of the rail; but it is to be remembered the number, tonnage, and speed of the trains, the character of the air and moisture, and the construction of the permanent way, and any particular circumstances will much influence results, and that no rule can be deduced. It is generally found that rails in a wet place in a tunnel wear quicker than when they are in a dry position.

In the St. Gothard Tunnel the water that percolates, or is formed, and trickles down the walls, is found to contain sulphuretted hydrogen, sulphur dioxide, ammonia, and carbon dioxide, besides other products from the smoke of the locomotives. In the middle of the tunnel the temperature is about 73° F. The corrosive and decaying influences are therefore bad, and for the telegraph cables it was found necessary to have an exceptionally strong protective coating. A St. Gothard Railway official report has declared that "the life of rails in long and ill-ventilated tunnels is scarcely one-third of their duration in the open air." The weight of the rails has, with a view to mitigate this effect, been increased from 74·6 lbs. per yard to 88·7 lbs. per yard. In a tunnel in France, 984 yards in length, which receives a little water by percolation, the rails are on beech sleepers, which rest upon gravel ballast. 230,000 trains, at a speed of about 19 miles per hour, had passed over them when they were withdrawn after 11½ years of service. The double-headed steel rails were 78 lbs. per yard, and 8·74 yards in length ; 10 sleepers to each rail. The original weight of a rail length was 682 lbs. After use it was reduced to 518 lbs., the loss being 164 lbs., or 18·76 lbs. per yard. The most important influence was the rolling of the wheel on the top table, combined with the rust which formed during the time elapsing between the last train in the evening and the first in the morning.

Abrasion and corrosion combine to reduce and wear away the area of the head of a rail, but the loss on the other portions of a rail is due almost entirely to corrosion. It has been found by careful experiment that by well covering wooden sleepers with clean ballast, the precipitation of acid gases to the iron rail flange is greatly prevented, and the passage of any acid water considerably hindered. At the eighth meeting of the German Railway Engineers' Association, held at Stuttgart, 1878, twenty-seven authorities were in favour of keeping the sleepers covered, and only seven to the contrary. It may be well to here note the recommendation of an Italian railway commission on permanent way, appointed in 1882, on ballast and sleepers. It was decided that though uncovered sleepers may be preferable in northern climates, it is necessary in Italy to completely embed them in ballast. It was considered by covering the sleepers, instead of the ballast being laid level with their upper surface, the timber was protected from the direct rays of the sun, sudden changes of temperature, moisture, dryness, and frost. Anything which prevents decay of the sleeper, or corrosive influences reaching it, will lessen corrosion of the rails.

Mr. H. Footner, M. Inst. C.E., of the London and North-Western Railway, examined * some steel rails of similar section in different districts, and, on the whole, the results seem to show that the character of the atmosphere much influences the life, and that steel rails laid in an agricultural district will not corrode so quickly as when they are laid in a railway through a manufacturing town and neighbourhood. In a district where there are many chemical or salt works, the wear and corrosion are very much more. Mr. Footner states, judging from an examination of some rusted rails, that

---

* 'Minutes of Proceedings,' Inst. C.E., vol. lxxxiv.

the rate of corrosion on the bright surface in contact with the wheel flanges has been approximately four or five times as rapid as upon the rest of the rail. The area of the head or bright surface in contact with the wheel flange is to the rest of the skin of the rail as one is to six approximately. The scale forming on the body of the rail affords a lodgment for dirt, and the two become a natural coating or protection, frequently shaken off, and as frequently renewed, but retarding oxidation, while on the other hand the passing wheel keeps the running surface clean and in a condition most favourable for oxidation. He concludes by saying that the relative amount of loss by oxidation depends upon the brightness of the surface, and the absolute amount upon the purity of the atmosphere. It also appeared their abrasion by an unbreaked wheel is small compared with the amount of loss by corrosion when the atmosphere is impure, and that all bright surfaces exposed to moisture similarly suffer, such as railway tires, &c., and lose by corrosion an amount hitherto scarcely suspected.

To wash the rails in the Hauenstein Tunnel, the Swiss Central Railway officials used hot water issuing at a high pressure, for water at a low pressure was found to diminish the adhesion. Slipping of wheels, which was very frequent, consequently now very rarely occurs, and the wear of the rails and tires is found to be much diminished and sand is seldom required. The effect of this intermittent hot water flushing at a high pressure on the metal has not yet been recorded, and it is not easy to foretell, for although the rail may be so cleansed as to free it from active corrosive films, the water may find its way through the pores to the metal, and as a rule, water of the same character has a greater corrosive influence when hot than when cold. Comparative experiments alone can determine the relative rates of corrosion, for although the wear of rails and tires is no doubt reduced very considerably by a free use of water upon them, and the tractive force required lessened, still dampness increases corrosion.

From a summary of reports and opinions, founded on results, the corrosion of metal sleepers on European railways does not appear to be a source of anxiety. Galvanising, oxidation by steam, a tar or other preservative coating, being seldom considered necessary, except when the sleepers are stacked for renewals on the side of the line, or are to be used in damp tunnels or cuttings, or in a locality where special corrosive influences exist. Then they are either coated with a tar preparation, or paint, as also if they have to be shipped. In northern climates, in a well-drained formation and permanent way under traffic, it is considered corrosion in metal sleepers need not generally be feared as being serious, but if they are stacked and not in use it will be. In fact, experience seems to indicate if a railway is used, corrosion will be inappreciable in northern climates, but if there is little traffic over such sleepers it will soon become important. In tropical climates the corrosion is much increased. Steel in sleepers has been found in some places to rust faster than iron, particularly when it is exposed to moisture.

If metallic sleepers work through the ballast to a clay formation, in wet weather, oozing of mud will occur, and the whole road-bed may become mud when wet, and when dry will be dusty, the soil will crack and shrink, and corrosive conditions will be active. The nature of the ballast, among other things, affects the rate of corrosion; thus, fang-bolts buried in clinker ballast are found to corrode rapidly, and it is the same with any iron that comes in contact with it. It is to be remembered all metallic sleepers, and necessarily their fastenings, which are laid as it were upon the surface of the ballast, and are so packed, are exposed to atmospheric, and therefore corrosive influences more than wooden sleepers, whether their tops are covered with ballast or not.

In temperate climates, wrought iron or steel sleepers may not seriously corrode during their serviceable life in the permanent way, but in hot, damp, or tropical countries, thin wrought iron and steel appear to corrode very rapidly as compared with cast iron, although there may be a change in the composition of the latter metal. Where timber is quickly attacked by the white ant, some kind of metallic sleeper, notwithstanding the defects inherent in them, may have to be adopted, unless the wooden sleepers are very carefully and effectively protected from their attacks. The advantages of metallic sleepers, apart altogether from the question of corrosion, have yet to be demonstrated, and, except under peculiar circumstances, their present employment hardly appears to be justified by general results, although here and there they have been approved and used on considerable mileages ; however, much depends on their form, quality of the metal, nature of the ballast, traffic, form of fastenings, and other circumstances. So far as simple corrosion is concerned, from a perusal of reports from all parts of the world, in northern or temperate climates the corrosion, with ordinary precautions, need not be feared when they are in use in the permanent way. The best ballast for them appears to be either sand or gravel, and they should be well covered with it, but sand is not desirable ballast, because it does not afford a rigid bed, it is liable to erosion, and to be blown away, and another objection to it is that it insinuates itself into the bearings and machinery, and causes frequent repairs, and a marked deterioration of the rolling stock and metallic parts with which it may come in contact.

Rolled steel plate sleepers, $\frac{3}{8}$ of an inch in thickness at the centre and $\frac{1}{4}$ inch at the edges, on the Indian State Railways, it has been stated, are not found to rust much, except when the ballast is impregnated with saltpetre and known actively corrosive substances. The sleepers are dipped, at a temperature of 300° Fahr., in a boiling solution con-sisting of 3 parts coal tar and 1 part tar oil, which forms a good hard coating, the solution being kept boiling by the passage of the hot plates through it. Mr. H. F. Bamber stated that it is very essential the sleeper should not be so hot at the time of dipping as to cause the solution to give off thick suffocating fumes, for the coating then formed flies off like coal dust, and is useless for preserving the metal sleeper from rust. After the execution under his superintendence of large quantities of steel sleepers, he found that they should in no case be dipped in tar at such a temperature as to give rise to thick yellow fumes. The lengths of metal sleepers tried in France, when they have been properly coated with coal-tar, it has been found, do not oxidise to any appreciable extent.

Iron longitudinal sleepers laid in 1864 on the Brunswick Railways, it is said, were found to be but slightly affected by rust 18 years after; and those on the Bergisch-Märkisch line, in badly-drained ballast, after being eight years in use, were found not to be more affected by rust than the rails. Anything from 20 to 50 years has been estimated as the life of a metal sleeper, but the nature of the traffic, quality of the metal, design, thickness, and strength of the sleeper, nature of the ballast, bearing area upon it, climate, absence of special corrosive influences, and care in maintenance, cause any rules to be untrustworthy, except all the conditions are exactly similar. An approved anti-corrosive coating for the metal sleepers, used on the Algerian Railways, is to temper them with coal-tar. It has been used with success for some 20 years.

A disadvantageous feature connected with metallic sleepers is, that the threads of the bolts used for attaching the rails to metal sleepers become rusted, and there is difficulty in keeping them tight. However, " after many years' trial, the Netherlands State Railway decided to keep the screw bolts as a means of attachment between the rail and sleeper." In 1865, the rails were fastened by four bolts 0·42 inch diameter, to each sleeper. In 1883, for the first time, it was necessary to renew 2000 of these 40,000 bolts, and the remainder were still in the permanent way in 1885.

No vegetation should be allowed to grow upon any deposit that may have formed on any metal. When iron or steel arrives at a port in the Tropics it should be at once taken to the site, as it is found oxidation is much more rapid near the coast than inland, unless there are localities in the interior of the country exceptionally exposed to corrosive influences. In tropical climates, iron quickly decays if not thoroughly protected, and, if buried in the earth, the rate of corrosion will vary according as the soil contains active corrosive agents or not. The great heat of the sun, warm damp atmosphere, and increased vegetable life, generate acids which quickly corrode iron.

In consequence of the small dimensions, sectional area, exposed position, and numerous connections of the parts of most iron roofs, corrosion will have a serious effect on their strength, and as there are many metallic roofs—such as those over railway stations and goods sheds, riding schools, drill and public halls, &c.—of various kinds, careful periodical inspection should be made to ascertain the magnitude of the corrosion, as many are peculiarly subject to corrosive influences.

In the 'Gesundheits-Ingenieur,' 1888, particulars are given of some experiments made by the Bonifacias Colliery Co., at Kray, to ascertain the best metal to withstand exposure to the acid waters of the pits. Tests extended for $6\frac{1}{2}$ months with wrought iron, steel, and "delta" metal (an alloy of copper, zinc and iron). Test pieces were $0 \cdot 79$ by $0 \cdot 79$ by $0 \cdot 75$ of an inch. They were suspended in the water, and were carefully weighed before and after, and the percentages of loss of weight were found to be respectively $45 \cdot 9$ for wrought-iron, $45 \cdot 45$ for steel, and but $1 \cdot 2$ per cent. for "delta" metal. The company therefore adopted "delta" metal for their underground machinery. These experiments can hardly be considered conclusive as regards corrosion, for a simple weight test is insufficient; and a chemical analysis, before and after testing, is necessary to enable any reliable result to be attained, although the metal may be particularly anti-corrosive. Still, the behaviour of some alloys is very varied and often peculiar; and, again, it is absolutely necessary the tests are conducted so that there are no interfering conditions, such as galvanic action, &c.

G

To prevent chemical action caused by electrical currents, the wrought-iron bars forming the skeleton of the statue of Liberty on Bedloe's Island, New York Harbour, and to which the copper skin of the statue is fixed, were first painted with shellac, and then enveloped in asbestos fibre, although it has been remarked they may eventually be found to be of little use should electrical currents be set up.

Some few years ago, in Paris, bare copper mains were used for the underground electric light wires. The conduits were made of cement concrete, and were considered to be water-tight, but the copper became covered with verdigris, and so reduced in thickness, that fears were entertained it might become so corroded as to be the cause of overheating with the normal currents. Earthenware troughs were found to be not waterproof, and, by a complicated electrolytic process, explosions were considered not unlikely to be brought about. One such occurred, although not a serious one, and gas was proved to have had nothing to do with it. Examination of the conduit showed that carbonate of soda and caustic soda were present, and the conductors were thickly covered with verdigris and copper chloride. The cause of the explosion was considered to be that the explosive gaseous mixture contained oxygen, hydrogen, and chlorine, the latter gas being due to the presence of chloride of sodium in the water which filtered into the earthenware conduit, the roads above having been strewn with salt to melt the snow, and caused electrolysis. The conductors are, generally, now covered with a bituminous compound to prevent electrolytic action, and earthenware troughs have been abandoned.*

In the Ferrara district, in North Italy, notwithstanding that the quality of the iron was good, the pipes in some artesian wells, 8½ to 11 inches in diameter, and ½ inch in thickness, were badly corroded in six months, a thin outer skin only remaining. On the other hand, the Modanese tube wells, which are generally of small depth, do not reach

---

* For further particulars, see *L'Electricien*, 1892.

sea level, and their water is from gravelly streams almost entirely free from iron. They appear to be quite free from corrosion. In the Ferrara wells, it was considered, two causes brought about the destruction of the pipes, one mechanical and the other chemical, viz., the friction of sandy and gravelly particles wearing away the skin, and, the water then being able to set free the iron, slowly dissolves chemically the interior core till it reaches the outer face. Whichever side is the lower, the sand will affect the most by gravitation. Wooden pipes have been, therefore, used in the Galara district, and, it is considered, long-continued experiment is required to solve the problem created, viz., the manner in which to prevent the rapid wearing away and corrosion of the tube wells.

At the Southampton Waterworks, Mr. W. Matthews, M. Inst. C.E.,* has stated, in the apparatus used for softening the water obtained from chalk wells and borings, "All the bright work on the filters and lime cylinders is, so far as possible, nickel-plated, as the great amount of condensation produced by the proximity of tanks, containing water at a low temperature, would otherwise render it almost impossible to keep the fittings free from rust."

At some of the soda nitrate works, where, in the water, salt and alkaline matters are present, the boilers have to be cleaned frequently, and at intervals of not more than two months. Raw nitrate of soda is a mineral deposit, which is considered to have been formed by decomposing animal and vegetable matters becoming acted upon by the salts present in sea water. Such soil or substance is actively corrosive.

In the case of a cylinder bridge-pier erected on the Bombay, Baroda, and Central India Railway, it was thought the rapid corrosion of the bolts was caused by the proximity of the sea, as the water was brackish, and, therefore, the action was much more rapid than in fresh water.

* 'Minutes of Proceedings,' Inst. C.E., vol. cviii.

CHAPTER VII.

THE CORROSIVE INFLUENCE OF SOILS, VEGETATION, SITUATION,
CLIMATE, RAINFALL AND WATER.

CORROSION of iron or steel buried in the ground will vary,
according to the nature of the soil and its character, the
porosity or imperviousness greatly influencing the relative
powers of corrosibility. Although an iron column or base
plate is in the soil, it cannot be regarded as altogether free
from air or fresh supplies of moisture, and other decaying
agents. When water enters the pores of any soil it displaces
the air upwards, and the liquid previously present downwards; hence, for a certain depth, varying with the composition of the earth, the metal may be severely tried, as
corrosive influences will be frequently renewed, and if such
water is derived from manured fields, sewage, or irrigation,
or from peaty soils in which humic acid is present, the
corrosive effect will be much increased. Those who have
made a life study of the question, and have conducted numerous experiments, state that when water has reached a distance
of even ten feet from the surface, it is doubtful whether it
has got entirely beyond all atmospheric influences, and may
not again be drawn up within the limits of superficial
evaporation. Hence the portions of a metallic structure
which are buried in the earth may be subject to severe
corrosive influences, depending upon the nature of the soil.
The amount of percolation is found to be governed more by
the time at which the rain falls, and the manner in which
it is distributed, than on its actual quantity.

The absorbing power of soils has also to be considered, for dry loamy earths imbibe more water than sandy soils, but the percolation may not be so great. The density, looseness, and open nature of the earth have also to be considered, and much affect the absorption, for particles in suspension, smaller than the interstices in the soil, and lighter than the water, pass through or into it till the earth becomes waterlogged, or the water accumulates upon an impervious stratum. Any other transferred matter is retained by the soil. Barren soil does not absorb so much urea as that covered by vegetation, but, in the latter case, all the urea in any sewage water is not abstracted, and, therefore, some remains to exert corrosive influences. The effluent of sewage-irrigation water contains some ammonia and chlorine, &c., and these have a decidedly active corrosive influence. The effect upon the paint or composition covering the metal has also to be considered, and this is referred to in the Second Part of this book, relating to Fouling and Corrosion in Submerged Structures and Ships, and their prevention by paint, &c.

Earths absorb water, or allow it to percolate. Sand, all loamy soils, gravel, and almost all earths having fine particles not compacted, readily admit water, if they do not absorb it. Gravel on chalk is considered to make a dry foundation, and to conduce to dry air, especially if the localities be at a considerable elevation above the sea, and are fully exposed to the sun. Blue clay, compact gravel, which is sometimes almost of the nature of rough weak concrete, and most rocks consisting of close even grains, and if not fissured, are of an impervious character; however, clays have moisture in them, and, it may be said, earth, or even rock, upon which metallic engineering structures have to be founded. In deep foundations, which may reach to a soil known to be not of a corrosive nature, rain water may percolate through the upper strata, or down fissures, and so corrosive influences may be conveyed to the lower beds. This action has been beautifully described as follows :—

"I see the leaning strata, artful rang'd ;
The gaping fissures to receive the rains,
The melting snows, and ever-dripping fogs.
Strew'd bibulous above I see the sands,
The pebbly gravel next, the layers then
Of mingled moulds, of more retentive earths,
The gutter'd rocks, and mazy-running clefts ;
That, while the stealing moisture they transmit,
Retard its motion, and forbid its waste.
Beneath th' incessant weeping of these drains
I see the rocky syphons stretch'd immense,
The mighty reservoirs, of harden'd chalk,
Or stiff compacted clay, capacious form'd.
O'erflowing thence, the congregated stores,
The crystal treasures of the liquid world,
Through tho stirr'd sands a bubbling passage burst,
And swelling out, around the middle steep,
Or from the bottoms of the bosom'd hills,
In pure effusion flow. United, thus,'
Th' exhaling sun, the vapour-burden'd air,
The gelid mountains, that to rain condens'd
These vapours, in continual current draw,
And send them, o'er the fair divided earth,
In bounteous rivers to the deep again,
A social commerce hold, and firm support
The full adjusted harmony of things."

At the point where a porous soil overlies an impervious there will be dampness, even if the water flows away. The word imperviousness is here only used by way of comparison, and not to indicate absolute imperviousness. Earth that has been excavated, deposited, or turned over, is more or less aerated and porous, and therefore is favourable to the admission of dampness, to corrosive influences, and the growth of vegetation. It is well to remember that made or polluted earth, to a few feet below the ground level, contains many organisms and impurities which exist to a very much less extent below any such depth. Experiments have shown that after ten feet or so below the ground, many exist in hardly appreciable number or quantity, and therefore increased corrosion may be expected at or near the surface.

Bacteria, however, have been found in wells, and it is thought were introduced into the subsoil during the sinking operations, and that they found a congenial soil for development.

Soils neither equally acquire heat, retain or radiate it, nor so absorb moisture. Differences of polish of a metal, its roughness or smoothness, and colour when painted, will also affect the radiation and absorption of heat. Clayey or marshy grounds cause a diminution of temperature, and, in warm humid climates, are unhealthy and actively corrosive. Light, calcareous, and stony soils, have a much healthier effect. The saline nature of a soil causes it to be decidedly corrosive.

Structures erected in cultivated soils will be subject to much more active corrosive influences than in barren ground, because the former are comparatively loose and porous, and freely radiate in the night the heat they absorbed in the day; therefore they become cooler, and hence they condense the vapour of the passing air, and it falls in dew. The effect of the sun is to harden earth, consequent upon the evaporation of moisture, and to bring the particles closer together, and make the mass more compact. If this were not so, heat and drought would penetrate the soil and kill all vegetation. In this respect heat retards corrosion. At what seasons of the year may corroding influences be usually considered to be most active? In autumn and winter. Why? Because the capacity of the air for holding water decreases after the summer season is past, and vapour is consequently deposited upon everything with which the air has contact; hence the necessity of preventing the surface of the metal being exposed to air and water, and the value of even simple grease as a protecting covering to prevent the humidity of the air reaching the metallic surface. For the same reason, in a damp climate corrosion will be much more rapid than in an arid district and one having a drying atmosphere and almost continual sunshine or constantly cold climate; in fact, it might be said, in comparing districts or countries, that wherever vegetation is least, corrosion will be reduced to a minimum—a damp tropical condition, vhich quickly produces vegetable life, whether mould or mose, or in higher forms, being the worst. In Great Britain, the rate of evaporation, of course, is most during summer, and the amount of humidity in suspension is usually greatest the last and first two months of the year, but not always so. However, it may be considered there is most *moisture*, not rain, in the atmosphere in the months which generally have the lower temperatures. Evaporation of moisture in earth being most in the summer months, it is favourable to the production of salts in the earth, which have a corrosive influence.

Although water may not percolate a clay soil to any considerable depth, cracks and fissures occur in .t in very dry weather, down which air and moisture penetrate. A clayey subsoil is also subject to bulging in cold weather, so that it cannot be said ironwork bedded in it only undergoes the corrosive influences that the soil possesses in solid mass. In sandy soils, on land reclaimed from the sea, the salt water soon loses its deleterious action on plant and vegetable life, but clay soils do so very slowly, and may require two years or more before they can be sown. Hence, clay soils impregnated with sea water will have a more active corrosive effect than sandy earths similarly situated. Brackish ooze, and silty clay, and made soil, are actively corrosive; brackish sand and shingle are less so.

Aluminous or clay soils consist of silicates of alumina, mixed with more or less sand, and often with lime, and their colour is considered to be due to iron. They pass into clay loams by greater admixture of sands. This, with 30 to 40 per cent. of sand they may be called clay loams; with 40 to 70 per cent., true loams, or loamy soils; from 90 per cent., or more, sandy soils. Clay loams become marly when they have from 5 to 20 per cent. of lime in them, and if they have over 20 per cent. they are called calcareous soils. Alluvial soils often consist of fine mud, and contain mineral constituents and some organic matter, animal and vegetable. Diluvial soils are more usually stony and barren, and consist of large gravel, boulders, stones and clay. The mere chemical composition of a soil will not indicate its corrosive influence, although necessary to its accurate determination, for the earth, from local causes, may be impregnated with substances and liquids almost unknown in the pure soil; still, to know the geological formation, with a due regard to any special local corrosive influence, will enable the probable corrosive effect to be judged.

Insoluble silicious matter forms in clean sand about 90 per cent. of it, the remaining 10 per cent. being oxide of iron, alumina, carbonate of lime, and various other ingredients, according to the locality. Sandy soils vary greatly; for instance, the sandy downs along the coast of the North Sea are found to be less contaminated than other soils, and they have very considerable oxidising power. The oxidation or nitrification of water, by percolation through sand, destroys organic matter contained in it. Sand downs decidedly attract atmospheric vapour not precipitated in the form of rain. This is not due to capillarity, i.e., the power which very minute tubes possess of causing a liquid to rise in them above its level, as, after long droughts, at a slight depth on the hill tops, sand is found to be damp, although the lower surfaces are dry to greater depths at such times.

A dry sandy soil is warmer than a wet marshy soil subject to great evaporation, and the latter has a corrosive tendency. Peaty soils are chiefly composed of marsh and water plants and moss, the layers having successively dried. Consequent upon the action of humic acid, they become changed, and are actively corrosive. The mists, which are of frequent occurrence over marshes and rivers at night, also have a corrosive effect, and are caused by the air over them being almost always near saturation. The least fall in temperature causes the air to release some of its moisture in the form of dew or mist. Fogs are caused by the condensation of the vapour of the air, when near the point of saturation, close to the surface of the earth, whereas rain falls from a considerable height.

Soil is sometimes composed of alternations of gravel and sand. The gravel may be consolidated by the admixture of carbonate of lime from limestone pebbles. Light loam usually contains a large proportion of micaceous particles. Some arenaceous limestone, on analysis, was found to be composed as follows: clay, $5 \cdot 60$; fine sand, $9 \cdot 30$; oxide of iron, $2 \cdot 00$; carbonate of lime, $80 \cdot 00$; carbonate of magnesia, $1 \cdot 90$; moisture and loss, $1 \cdot 20$: total 100. Sulphur is found in mica slate, limestone, beds of gypsum, sandstone, in alluvium, &c. Soils obtain carbonic acid from the air from which it descends by rain. It is also abundantly present in all decomposed vegetable and animal matters. Limestone, chalk, and all calcareous stones, have carbonic acid in them in a solid state, and the presence of any acid sets it free.

The soil of Egypt, some analyses in reports made to the French ministry of agriculture have shown, consists of about 45 per cent. of silica and 53 per cent. of clay, and in general is composed of silica, alumina and oxide of iron, with small quantities of phosphoric acid, and not more than from 4 to 9 per cent. of organic matter; and a larger proportion of chloride of sodium is found on approaching the Mediterranean, attaining 4 per cent. in the middle of the delta. The Nile mud contains, on an average, at its high and low stages respectively, 55 and 58 per cent. of silica, 20 and 23 per cent. of alumina and oxide of iron, and 15 and 10 per cent. of organic matters, together with small quantities, $0 \cdot 66$ to 3 per cent., of phosphoric acid, lime, magnesia, potash, soda, and carbonic acid.

The value of chemical analysis of an earth or mud can be judged from the preceding examples in any reliable determination of the presence of corrosive substances in it.

The alluvial matter in river water is either deposited on the shores, increasing as the course nears the sea or mouth, or the current is sufficiently strong to carry it away. If not, salt marshes will be created, which is usually the case in estuaries. Such waters and earths have an active corrosive influence on metallic structures, and not infrequently in other situations various corrosive salts are found in them. The soil of forests is generally covered with long damp grass, rotting leaves, brushwood, &c., and there are places full of or holding stagnant water. Low-lying and damp situations also give a maximum of atmospheric pressure.

Climate, soil, humidity, light, elevation above the sea, &c., cause plants to grow in some localities and not in others. As regards temperature, it is the degree of heat, and the duration of the summer months, that is the chief agent in limiting the range of vegetable life, and not the mean annual highest or lowest temperature that may prevail. At the sides of a pond, or sheet of water, leaves and débris will be found, unless driven by wind, because the shore attracts the leaves, &c. Vegetation is also, on the margin of a river, more abundant and strong than in an open field, because the porous earth on the bank draws, by capillary attraction, water to the roots of the plants. Consideration of the causes of plant life will enable its probable corrosive effect to be ascertained. The green colour of plants is given by light, the intensity increasing with the brilliancy of the light. Heat and moisture are the other indispensable requirements for the vegetation of plants. One of the chief causes of the remarkable growth of tropical vegetation is the considerable quantity of water contained in a tropical atmosphere in the condition of transparent gas. It is known to be proportional to the heat, and to be not affected by what is known as a clear sky. This invisible water is absorbed by the plants, and taken up by their large leaves, and so they luxuriate to such an extent as to be of marvellous growth, whereas in our temperate climate, without rain vegetation would be parched up and perish. This beneficently bestowed property of tropical regions actively produces corrosive and fouling influences.

In a valley in a mountainous country corrosion may not equally proceed on both sides of it, for there will be a sunny side and a more or less sunless side, and more moisture necessarily on one than the other. The direction of the valley or gorge will, therefore, govern the rate of corrosion, and the sunny side is almost always the one to be preferred for the erection of metallic structures, if preference can be given, simply from the point of view of facility and construction and durability. The character of the vegetation will indicate the warmth of climate. For instance, in the Alps of the Valais, on one side may be seen the vine growing, on the other thick ice. The slope of the country, and the aspect it presents to the sun's course, causes this difference, the side on which the vine grows being exposed to the direct rays of the sun. On the southern terraces of the Himalayan slope, the pine-apple, cotton tree, mango, &c., grow, even at an altitude of 5000 feet; this tropical vegetation then gives place to the plants of a temperate region, and, higher still, to alpine forms of vegetation. The power of the sun in heating the soil varies according to the angle at which the sun's rays strike the ground. When the sun is at an angle of 45° above the horizon, the sun's rays will fall perpendicularly on a hill facing the south having an equal slope. The level land below will receive them at an angle of 45°. If the northern slope of a hill inclines at an angle of 45°, the rays will be parallel to the surface, and their heating effect will be very little. If the northern slope be steeper than 45° it will be in the shade. A S.S.W. or S.W. aspect is the warmest in Great Britain and the Continent, and a N.N.E. or N.E. the coldest, unless special local circumstances cause a modification of the general condition ; therefore, corrosive influences will vary, and while one side of a hill may be exposed to them, the other may only be to a much less extent.

It is known that some lakes and many rivers present an exceptional climate. Such peculiarities are usually the result of protection by mountains and hills from north and other cold winds, and from their being exposed to the sun. Also, near some rivers, what are called " river clouds " occur at certain seasons, and this is considered due to the temperature being much more constant than further inland, which is quickly cooled by radiation. Vegetation which does not exist inland often thrives on the borders of lakes, and so just around the Lombardy lakes vegetation flourishes that is found usually only in much more southern and warmer places. It is the same at Lake Geneva, the north side of the lake, exposed to the sun's rays, and sheltered from north winds, being more favoured than the south. It was found by experiment that, with the sun at an altitude of 30°, the reflected heat was nothing : and 60 to 70 per cent. of the incident heat between about 4° 30′ and 3° 30′. The reflected heat at 16° 30′ was about 20 to 30 per cent. of the incident heat. The north and south bank of a river stretching east and west, and sheltered by mountains or hills, will usually show different plants growing on one side to the other, the north side having those which grow in the warmer regions of the earth, for the north side faces the south. These phenomena it is well to remember, as corrosion is not likely to be the same on both sides of a river or mountain under such circumstances. On the north side, having a southern aspect, there would be much more sun and drying of the surface.

In connection with the special subjects of this chapter, it is advisable to ascertain the annual quantity of rainfall, and months in which it falls ; whether mists, fogs, dews, &c., are frequent, and the atmosphere usually in a damp, warm condition ; the latitude and elevation of the site of a metallic structure above the sea-level ; the location of the site as regards the sea ; form and nature of the land, whether mountainous, swampy or wooded ; its position with respect to the sun's course, especially in a valley, as on one side the sun may shine, but not on both, and perhaps never on either ; the character of the sea, and waters of rivers, &c., rain, and the nature of the soil ; the prevailing winds, remembering that those countries where the winds are the most variable are the most cloudy, as Great Britain ; the rate of growth, and species of vegetation, also of mollusca, and all life that aids corrosion, decay, and fouling ; the effect of these influences on the metal, and on any paint or composition used to cover it or supposed to protect it ; and also, whether, as in screw or other piles or ironwork, the paint is likely to be removed or injured in screwing or during erection.

The rainfall, or moist character of the air, which is governed, among other causes, by the situation and configuration of the land, and its impurity or freedom from corrosive agents, will vary in almost every district on the known surface of the earth. Natural water being impure, because of its contact with the earth and the atmosphere, it is the character and intensity of the impurities in it as regards corrosive influences that have to be here regarded. The quantity varies greatly : thus, at Bergen, the commercial capital of western Norway, the rainfall is excessive, nearly four times that of the east coast of England, and amounts to about 88 inches of rain, the heaviest of any city in Europe, and more than that of many places situated in the tropics. The reason is, clouds are driven forward by south-westerly winds into the Norwegian fiords, until they are arrested by the mountains, and accumulate, then rain is caused.

The proximity of mountains attracts vapours and condenses them, and hence heavy and continuous rain occurs. As a rule there is more rain in islands and on sea-coasts than in inland districts or places far removed from the sea—as the broad plains of Asia, Africa, Siberia, some parts of Eastern Europe, and America—and in mountainous places than in flat countries, and in the tropics than in more temperate zones, with a few exceptions, such as Bergen. In general, the rainfall becomes less with the increase of the latitude, but mountains or numerous hills will cause a few exceptions, as at Bergen, Keswick, Kendal, &c.

The causes of rain and deposition of moisture are so well known that there is no occasion to here refer to them, except so far as is absolutely necessary. The quantity of rainfall or moisture retained must vary according to the form of the land, as witnessed in mountainous countries where the water may at once flow away, or be retained as in marshes or flat localities. The character of the surface soils and subsoils will much affect the corrosive activity of water flowing over or through them. In the north and south zones of variable rains, i. e., from between about latitude 20° N. and the Arctic circle, and 20° S. and the Antarctic circle; the zones of periodical rains, say, from between latitude 20° N. and 20° S.; the zone of almost constant precipitation, situate about 3° to 10° N., extending from South America and reaching to Africa; and in the rainless and riverless districts in the north of Africa and a few other places; corrosive influences must necessarily vary, and it is, therefore, indispensable that the characteristics of the country in which a metallic structure has to be erected be duly considered.

Dr. Angus Smith has shown that there are innumerable solid particles in the air, consisting of common salt, sulphate of soda, nitrate of ammonia, and sometimes lime-salts and iron, as well as phosphates, iodides, and organic substances, given off from animals and vegetables and living things, and probably a little of nearly everything at times. The percentage of oxygen in the air varies from all but 21 per cent. to 20·40. After rain, the atmosphere contains a larger percentage of oxygen than it did previously, and there is at the same time a diminution of carbonic acid. Rain brings down the "atmospheric dust," and carbonic acid and other gases, by the mechanical and chemical actions of the falling water. In rain waters, chlorides, sulphates, and ammonia have been found. Hydrochloric acid in the air is due to the salt found near the sea, and in the neighbourhood of towns, where it is one of the products of coal burning; but rain water near the sea is not acid, but that taken in towns is so because it contains sulphur compounds, and the latter also become present where decaying animal and vegetable matter are. Sulphuric acid is a measure of decomposition, being a part of the oxidised sewage of the air. The acidity of rain water is also a measure of the impurity of the atmosphere. Dr. Angus Smith found that, taking the inland country parts of England as 0, Glasgow figures 109, Manchester and Liverpool 83, and London 28. The comparative amounts of combined ammonia are shown by Valentia 1, Glasgow 50,

Liverpool 30, Manchester 36. Hydrochloric acid : Blackpool
100, London 320, Underground Railway 974. These being
comparative figures, as are the following :—Sulphuric acid
(anhydrous) : Blackpool 100, London 352, Underground
Railway 1554. Ammonia and albuminoid ammonia : Innellan
on Firth of Clyde 100, London 117 and 108, Glasgow 150
and 221, and the Underground Railway 138 and 271. Single
experimental results should, however, not be regarded as
absolutely conclusive, but the mean of a number be found
before deciding upon the amount of impurity in the air of
any particular place. It may be useful to remember these
relative quantities when considering the probable corrosive
influences to be encountered.

The rainfall of manufacturing towns, and of cities and
towns, has a much more corrosive effect than that of rain
as it falls in the open country or mountainous districts,
because of its impregnation with the air; but comparatively
pure rain water may not be so after percolating the earth
for a little distance, because of the corrosive effect of manures
and substances in it in well-cultivated land. It may be said,
generally, the impurities of the air are indicated by an
analysis of the rain water of the locality, and that in dense
manufacturing districts, or a copper-smelting and chemical-
producing town, such as Widnes on the Mersey, acids are
present which can hardly be said to exist in more favoured
places. Many acids have such a strong affinity for water,
that it is very difficult to entirely separate them from it.

Water has very great solvent powers, and evaporates at
all temperatures to which engineering structures are subject.
Rain waters, and dews to a lesser extent, in country districts,
in falling to the earth, take up the gases, oxygen, nitrogen
and carbonic acid, a little carbonate of ammonia, nitric acid,
&c., including any particles which may be floating in the air.
In towns and manufacturing districts, in addition, sulphurous acid, sulphuretted hydrogen, &c.; and if rain water runs
from a roof or down pipes, the impurities in it will be
various. In considering the corrosion caused by rainfall,

it should not be forgotten that in many tropical countries the rain descends in torrents for many hours, and in quantity even as great as about one inch in an hour, and also very violently, so much so as to cause districts to be flooded in a short time. The effect of such rain is to very severely try the paint on a metallic structure, as the constant beating of the rain against its surface will wear away any but the most tenacious and hard coating. An instance of the destructive effects of tropical rains in damp situations may be mentioned. In the Western Himalayas, Major Browne, R.E., has stated, "Retaining walls of what seemed most compact sandstone have suddenly collapsed, the underground courses have become dissolved into sand." Such walls, from the foundations to two feet above the ground line, are therefore built of granite boulders. This shows the active decaying force that may be at work upon any painted or coated surface of iron or steel, and the desirability of the metal being embedded'in impervious Portland cement concrete, or otherwise effectively protected.

It would be of no practical value to give analyses of the water of different seas and rivers, for they are subject to frequent changes and notable differences. The character of the water resulting from rain or floods will generally vary somewhat daily everywhere, and it is advisable to consider the kind of soil, whether argillaceous, calcareous, ferruginous, or of a special nature, that the water has flowed over or percolates through. In the sea, rivers and streams, consequent upon the ceaseless change of water, and consequent constant fresh supply of corrosive influences, iron or paint are in a perpetual state of siege, and very much more so than if the water was quiescent. Analysis will seldom show the composition of the waters of rivers to be the same for the total solid residue, combustible matter, the ash, sulphuric acid, chlorine and common salt, iron, ammonia, nitrogen, &c. The bed of a stream, in some cases, may be covered with weeds, entangled and dense, and replete with crustacea, mollusca, and the lower forms of life, and this becomes less

H

till at last the bed appears to be clean, and the water quite clear, although still full of animalcula. Fresh water algæ exist in large numbers and great variety in most water, and are very hardy and tenacious of life. They flourish even in the Arctic regions. The species are very numerous; some seem to live in waters containing sulphur, salt or iron, some in stagnant water, and others in the clearest and purest. Starch and globules of oil have been found to spring from them, and decay and taint the water, also a kind of jelly from diatoms. Especially in many warm climates, rivers are generally in a turbid condition, and the water never free from argillaceous substances, and, if exposed to the sun, the formation of fungus on the surface of the water will be more or less rapid, according to the location and special circumstances of the site of a structure, and will be greatest during the spring and summer season.

The composition of water varies greatly, and is affected by the nature of the ground it passes over or through. It has in it organic and inorganic compounds, and the inorganic compounds may be sodium chloride, carbonate, nitrate, sulphate, sulphate and carbonate of calcium, sulphate and carbonate of magnesium, silicic acid, alumina and oxide of iron, potash, chlorine, ammonia, phosphoric oxide, sulphuric acid, &c. The salts usually contained in hard water are carbonate of lime, chloride of sodium or common salt, sulphate of lime, sulphate and carbonate of magnesia, and sometimes carbonate and oxide of iron. Spring water contains a considerable quantity of carbonic acid, and is, therefore, pleasant to drink. The chlorides, sulphates, and nitrates of calcium and magnesium are easily dissolved and maintained in solution by water, but the carbonates of these elements can only be maintained in solution by an excess of carbonic acid in the form of bicarbonates. Mechanical impurities in water are removed by filtration, but chemical impurities may not be. A quick method of ascertaining whether there is vegetable and animal matter in water is by adding a little sulphuric acid to it, the water will then become black. Water charged with carbonic acid disintegrates, by continuous action, limestone, granite and hard rocks; and zinc and galvanised iron, which do not appreciably corrode in a clear and comparatively pure atmosphere, if exposed to the carbonic acid, which is present in the air of some manufacturing towns, and is also in abundance in some water, decay, and the galvanised coating becomes loose or detached, by corrosion of the surface of the iron underneath. How is the carbonic acid in water produced? By the presence of bicarbonate of lime, which is a salt composed of lime and carbonic acid, and is frequently held in solution by hard water, and chiefly causes it to be more agreeable to the taste than soft water; and hard water is made softer by exposure to the air, because not only do the mineral salts subside which cause it to be hard, but the carbonic acid of the water escapes into the air. A copious supply of water prevents the accumulation of carbonic acid by dissolving it, but it causes the water in which it is dissolved to be slightly acid.

An analysis by Dr. Lyon is here given of what may be called an actively-corrosive water, viz., that at the estuarine mouth of a river. The water of the Nerbudda River, in India, taken at a bridge-crossing of the Bombay, Baroda and Central India Railway, was found to contain per gallon—

	Grains.
Chloride of sodium	139·40
Chloride of magnesia	8·64
Sulphate of zinc	11·34
Sulphate of magnesium	7·84
Carbonate of magnesium	9·37
Silica	1·34
Organic matter	·56
	178·49

Suspended matter, 72·94. Specific gravity of clear water, 1·00206. Total solids, dried by evaporation at 280° F., 178·64.

H 2

When water is rich in lime, potash, and sodium chloride, with marked traces of nitrates and phosphoric acid, it is known pollution has taken place. There will be a quick growth of vegetable organisms, weeds and mud will be produced, and then corrosion and fouling may follow. Water containing ammonia, fœcal and excrementitious matter, and that impregnated with salt, is actively corrosive. Sewage water, in addition to other constituents, usually contains nitrogen, sulphuric acid, chlorine, phosphoric acid, potash, soda, lime, ammonia; the nitrogen being present in organic compounds and as ammonia, but in the effluent chiefly as ammonia and as nitrous and nitric acid, and only in small quantities in organic compounds. Nitric acid is frequently found in impure waters in considerable amount. The quantity of suspended matter in, and the character of sewage varies greatly according to the locality. It will also be affected by the waste effluent from breweries, factories, and by water from mines, peat bogs, chemical works, wire-drawing manufactories, where acid is used for pickling the wires, and by any which contains free acids and in which salts have been dissolved, such as chloride of magnesium; also that from dung-heaps and slaughter-houses. Water of this description is decidedly actively corrosive. The character of the drainage waters of towns or manufacturing districts is different on Sundays and holidays to that of the ordinary working days.

The first flow of a flood will generally bring down vegetable and organic matter, and then alluvial matter, as the floods, which are frequently helped by gales, increase; and so corrosive and fouling influences are seldom regular and of the same power. At one time, as in dry weather, the alluvial matter carried may be very small, but it will probably be large in wet weather and at flood time, and particularly so if a flood or gale occurs after a long drought. The character of the sand and mud and matter in suspension in river water can be sufficiently ascertained, but the material that rolls along the bed and at the sides cannot be so

nearly determined. The dissolved matter, or that in suspension, will be dependent upon the condition and nature of
the soil it has passed over or percolated through, and the
length of time taken to complete either of those operations.
The harder silicious rocks will give off but little mineral
matter in solution to water passing over them, or even rising
in fissures in them. But in the soft rocks, chalk, or calcareous
earth, marly and alluvial soil, and such soft earth, it may
be found in much increased percentage, either in solution or
in suspension. Waters which hold in solution much organic
matter quickly putrefy. Ponds, marshes, bog-land, most
stagnant waters, &c., are examples of this. The organic
matter becomes altered, and is in suspension or dissolution.
Anything which causes putrefaction must be considered as
an active corrosive agent to the metal and the paint. The
effect upon the latter is considered in the second part of this
book, on Fouling and Corrosion in Submerged Structures and
Ships, and their prevention by paint, &c.

Waters flowing over rocky or clean gravel, or sandy beds,
are obviously much purer and less corrosive than those which
have been contaminated with the surface earth of cultivated
land, or deposits that become in a state of decay or putrefaction from the decomposition of vegetable or animal matters.
The substances contained in solution in the waters of the
Nile do not materially differ from those found in many other
rivers, but their relative proportions appear to vary according to the stage of the river. Again, the amount of silt
deposited in reservoirs depends upon the catchment area,
rainfall, and the geological character of the soil. For
instance, in Algeria, in some reservoirs, it ranges from $0 \cdot 16$
to $1 \cdot 6$ cubic yards yearly per acre of catchment. The silt
hardens by time, and therefore it is easier to clear, being
looser, and to carry off, if done yearly or so. While water
flows underground its impurities are held in solution by the
presence of carbonic acid, but when the stream reaches the
open air its carbonic acid escapes, and these impurities, which
are especially carbonate of lime and iron, are precipitated on

various substances lying in the course of the stream, and cause petrifaction.

A structure exposed to petrifying spring water, or incrusting springs, i.e., those covering things with a crust of the mineral they hold in solution, obtained in trickling through the ground by dissolving some of the substances they encounter, as in springs from limestone mountains which incrust substances with lime or calcareous matter in a short time, brine springs with salt, sulphur, iron, mineral and medicinal springs, are also numerous, and most of these do not materially change, or have not changed since they were discovered, and may be considered as actively corrosive. The influences of light, air, evaporation, loss of temperature, absorption and escape of carbonic acid, cause them to deposit the mineral matter. Some mineral springs have a high temperature. In Italy, Mexico, and some parts of Spain and Portugal, they are numerous. Those in Great Britain are so well known that they need not be specified. There is hardly a country which does not possess several. Chalybeate springs, or those containing iron, are very numerous, and deposit oxide of iron wherever they flow. Saline springs less frequently occur. Many of the chalybeate and saline springs are impregnated naturally with carbonic acid gas, which makes them exhilarating. Any metallic structure subject to any such waters requires to be specially protected against corrosive influences, and the effect of the waters on the paint or composition used has to be duly considered, as any coating may soon be deleteriously affected.

It may here be well to refer to some experiments made by Professor August Wagner upon the rusting of iron under water at the ordinary temperature. He found that under distilled water, saturated with carbonic acid and air, iron rusts nearly twice as much as under water containing air only. If, instead of distilled water, spring water concentrated by evaporation is used, an essentially different result is obtained; in this case, iron rusts in water containing only air more strongly than in water containing air and carbonic acid.

And it was generally found that iron under water containing air, but free from carbonic acid, with the presence of salts, except alkaline reagents, rusted much more quickly than in pure water. In water containing chloride of barium and chloride of calcium, saturated with air free from carbonic acid, rusting is very energetic, but less striking with air and carbonic acid. Under water containing common salt and chloride of potassium, as well as under water containing ammonia, rusting in the presence of air and carbonic acid proceeds with the greatest energy. Under water containing chloride of magnesium, rusting occurs with air and carbonic acid more strongly than in the presence of air free from carbonic acid. Additions of oil greatly prevent rust in water containing only air, as well as in water containing air and carbonic acid. Alkaline reagents, as lime and soda, tend to prevent the rusting of iron. Both at the ordinary temperature, and at the boiling point, the presence in water of such dissolved chlorine combinations as chloride of magnesium, of ammonium, sodium, barium, potassium, and lime, is very destructive to the iron so soon as air is admitted. The addition of alkaline substances, such as lime and soda, to the feed-water of boilers prevents, it was found, rusting of the boiler plate.

Experience and experiments show that hot sea water acts more corrosively on boiler plates than hot fresh water, and that the less sea water there is mixed with the fresh, the less the corrosive influence; also, the less frequently either water is changed or renewed, and that, if the immersed metal is occasionally exposed to the atmosphere, and again immersed, corrosion is accelerated. Mr. Hering, M.I.C.E., with regard to the pollution of streams, draws the following inferences from the best data available in 1888 : " That rivers *not* to be used for domestic water supplies, but which must remain inoffensive to communities residing a few miles below the outfall, to be fit for all manufacturing purposes, and to sustain the life of fish, may receive the sewage from 1000 people for at least every 150 or 200 cubic feet minimum flow per minute." Any such water, however, will have a corrosive influence.

Lakes having no stream flowing through them are to be regarded as stagnant waters, but the rock on which they generally repose, and aeration of the water by wind, much lessens the effect of no flow, which would otherwise aid the growth of weeds and decomposing influences. Why? Because plants grow in stagnant water, and leaves, &c. become decomposed in it, and insects lay their eggs in the leaves and plants floating on the surface, the result being a gradually increasing swarm of worms and insects. There is usually more suspensory and corroding matter in river than in lake water, chiefly consequent upon the greater number of cities, towns and villages per acre of water area, and the more polluted nature of the supplies, and also because of the increased facilities offered in lakes for the deposition or precipitation of suspensory matter, owing to the comparative stillness of the water. The salt in the water at the mouths of rivers is due to admixture of oceanic brine with the outflowing fresh water, and at such places corrosion is likely to be severe, for brackish water acts more rapidly on iron than fresh water. Sea water flowing up a river will much increase the percentage of chlorine in it. The salt water so entering, it has been noticed, sinks to the bottom. This is owing to the different specific gravity of fresh and sea water, which, therefore, do not readily commingle, the sea water being separated from the fresh. Fresh water being lighter than sea water, it, as it were, floats and spreads on it until it becomes commingled. A current of salt water meeting fresh, will, at first, force itself under the fresh water. Also, water heavier than that into which it is discharged, as sewage, displaces the lighter water, and proceeds under it till intermingling occurs by deposit of suspended matter, or by other causes. Remembrance of this may indicate its possible relative corrosive influence, and also where it is likely to be active.

The viscosity, or frictional adhesion, of water to the surface of a structure will vary according to the character of the water, and nature of the material used in a structure. Although water, under any probable pressures occurring in practice, is not increased in viscosity by the force it is pressed against a structure, its corrosive effect is greater. Some recent German experiments have shown that a mixture of 25 per cent. of common salt with water increased its viscosity at high pressures and temperatures. The viscosity of concentrated aqueous solutions of sodium chloride and ammonium chloride are also increased approximately proportional to the pressure, and almost independently of temperature. It is well to bear in mind the influence of pressure and temperature upon the viscosity of fluids in endeavouring to ascertain the probabilities of corrosive effect.

In ordinary water, the dominant action upon iron is that of the oxygen of the air dissolved in water, for, although water is decomposed by contact with iron, it is chiefly the oxygen of water which causes the corrosion of iron, all water containing atmospheric air, the percentage varying. Water may act on any surface by percussion, by pressure, by reaction, and by chemical action. The chemical action of gases has to be resisted as effectually as the corrosive or solvent action of water. Mechanical agitation also facilitates solution. Experiments have shown that falling water exerts a pressure due to a column of water of a height equal to the fall. Wind pressure, however, may increase or even diminish it. All dripping water from projecting parts should be kept from a structure. Pressure of water has an effect on the painted surfaces of a structure, for it may find out any weak places, as it does when it presses against scales of iron rust. With a pressure of 1 lb. or so, or that equal to about a depth of 2 feet 4 inches on the square inch, they may readily resist percolation of water, but it may not be so in 20 to 25 feet of water, when the pressure is about 9 lbs. to 11 lbs. per square inch, and water may ooze through. Constant pressure from still water, i.e., without any impactive force, cannot have such an effect on the paint as that of waves, the pressure of which on the side of a structure has been recorded as equal to about 50 lbs. per square inch. Some experiments have shown that the corrosive effect of active water in the open sea is some seven times more that of the same sea water when kept still. The different effects of still and moving water upon the rate of corrosion may be shown by examples, namely, that corrosion, except under peculiar circumstances, usually increases externally in ships upwards from a few feet above the keel, the worst places being situated near the wind and water line, and a few feet down, that is, where the plates are alternately wet with the sea water, dry and exposed to the air. Experience and experiments have proved that in water constantly aerated, or changed by currents, or disturbed, such as the surface waters of seas and rivers, which continually supply fresh quantities of active corrosive agents, both iron and steel become corroded much quicker than in still water. That the constant wash of bilge water in ships, when the internal plates are not properly protected, soon corrodes and destroys the heads of rivets, and grooves the plates between the frames on the path the water takes, and that main pipes in which water is constantly in motion become quickly rusty if not covered with an anti-corrosive preservative; whereas in disused or seldom used branch pipes for firecocks, and houses to which the supply has been discontinued, although full of comparatively still water, and only occasionally subject to a flow, corrosion is comparatively insignificant. Here, also, may be mentioned that corrosion in boilers is usually the quickest in the stays at or about the water line, and in the steam space.

These facts show that the more the water is aerated the greater will be the corrosion, other conditions being similar, and that the constant *renewal* of the water practically replenishes the surface with a new and incessantly energetic corrosive fluid. It is important this should be remembered, for it indicates some parts in a metallic structure which are more liable to corrosion than others, and therefore require to be especially protected and examined. Pressure of water, such as that caused by wave action, will also aid corrosion, and is by no means to be disregarded in submerged or occasionally submerged structures, for if scale is formed on the iron underneath the protecting coating, or if a weak place occurs in the covering, pressure will at length cause a perforation of the paint or scale, and then water will percolate till it fills any unoccupied space, and will become compressed according to the head of water or the force of the waves, the result being that the scale or protecting coating is forced away. The scale, also, not having any elasticity to yield to the straining of a ship or submerged structure from the shock of the waves, cracks and becomes detached, and, even supposing the paint is sufficiently elastic to be unaffected by such sudden and repeated strain, the scale will ultimately cause the paint to crack, and then the pressure of the water will complete the disintegration, and the metal, or what is left of it, if in thin plates, will be bare and entirely unprotected, and be subject to influences peculiarly liable to destroy it. Various iron and steel structures subject to tidal immersion, both in sea and fresh water, which have come under the observation of the Author during the last quarter of a century, tend to show that in a properly protected and maintained bridge, landing, or promenade pier, the greatest corrosion is upwards from about low-water level, and that from a few feet below any wave action, although there is corrosion, it is much less than that above what may be called the line of still, or nearly still, water. In other words, in iron or steel work constantly submerged, corrosion proceeds more slowly than when it is alternately immersed and exposed to the atmosphere, especially if the water frequently or regularly changes from being fresh to brackish; the chief exception being in water impregnated with sewage, when the corrosion will be very active, and will continue so until a depth is reached in the ground to which the sewage water has not penetrated.

Iron does not appear to corrode in ice, but whether oxidation already commenced is arrested is doubtful, although it appears to be much less active; and so long as ice remains in water, and the temperature does not exceed 32° F. or cause wet ice, corrosion is not apparent, unless air is allowed to reach the metal. This is the case with ordinary hard water, such as London water. At about 60° or 70° F., however, corrosion becomes apparent. In an hour or two a little scum of greenish hue can be seen on the surface of the water, and in two or three hours it has a brownish colour at the bottom. After a few weeks, thin rust particles will be formed on the surface, and can generally be washed off with water, but if they are raised lumps, which are usually concentric, the plate will be found to be pitted, and serious corrosion has commenced, and loss of sectional area of the plate. Whether metals suffer from freezing of water in the pores has not been determined; however, the expansive power of water is very great, and it is well to bear in mind that it expands by heat as well as by cold. It expands from 40° F. at a higher temperature till it becomes sufficiently hot to be converted into steam, or so cold as to become ice, when, by its conversion into solid crystals, which do not fit so closely as the particles of water did, it expands.

Ocean currents, or marine rivers that flow from the high to the low beds of the sea, are believed to be promoted by differences in the density of the water, evaporation, rotation of the earth, character of the coast, rivers flowing into the sea, tides, winds, effects of climate, and other causes. Each will have its own corrosive influence, and it may vary considerably; for instance, sea water, by eroding the cliffs or lands in the vicinity of a structure, will probably deposit slime of a similar character upon it.

Experiments have shown that light penetrates both fresh and sea water to distances far beyond any upon which an engineering structure is likely to be erected, and that the actinic rays do, to at least 1000 feet in the open sea, and to about half that depth in lake waters, which are open all over their surface to the sun's rays; so that if light is necessary for the promotion of any fouling, and consequent corrosion, it is present.

CHAPTER VIII.

GALVANIC ACTION AND CORROSION.

The influence of galvanic action in the promotion of corrosion in iron or steel, whether caused by immediate contact or by mediate contact, as sea water, wire, &c., is important. The direction of the current of electricity has been ruled to be, that the current passes from the metal subject to the greatest chemical action to the liquid, thence to the sound metal, and so forward, and therefore the direction of electricity in voltaic arrangements may be known, and the relative designations of positive and negative can be allocated. The decomposing effects of galvanic electricity on water are marked, the oxygen being liberated at the positive end of the battery, the hydrogen at the other; and although energetic decomposition may be slow, consequent upon a bad conductor, such as water, being the conducting power, the latter becomes readily increased by the addition of a little acid or common salt, hence the accelerated decomposing effect of rain water in dense manufacturing districts, and of sea water on iron or steel under such conditions. It is well to bear in mind, in metallic structures, that voltaic action, as evinced by chemical decomposition, may ensue without metallic contact. Electrolysis, i.e., the decomposition of a compound substance by the action of electricity or galvanism, is now generally accepted as a fact.

In considering the effects of galvanic action, and the manner in which it chiefly arises in structures, it is impracticable to state every condition in which it may occur, for one of the highest practical authorities has shown that " in

one and the same piece of plate, galvanic action might come
into play very decidedly in promoting the corrosion of the
metal." Indirect or circuitous connection is sufficient to
produce it, 'for instance, if the metals be in some corrosive
liquid capable of corroding one of them. The importance of
uniformity in the composition and quality of iron or steel
has been previously referred to, and it may be taken that
uneven corrosion may occur because of electric action generated by oxide being crushed into the metal in the piles
during rolling. When this electric condition exists, the
effect is necessarily uneven in intensity, and generally
variable according to the position and quantity of oxide that
was on each pile of the bloom, or on the metal at the time of
rolling. There can be no question that the composition and
purity of the metal very much influences the decay or corrosion, but the desired purity must not cause difficulty in
working the metal, or reduce its tenacity or other powers.
The insertion of a new sheet into a boiler, consequent upon
a modification in the form, and not because of excessive or
uneven corrosion of a plate, has been known to set up
galvanic action so strongly that all the sheets became corroded except the new plate. This, by Dr. Cresson's experiments, was shown to be because of the different electro-
condition of the sheets, the purer iron corroding and protecting
that which contained the greatest amount of carbon, the
new sheet of iron being electro-negative to the old plates, its
position in the electro-chemical scale depending upon the
amount of carbon it contained. Neither a black or any
deposit should be allowed to remain on a metallic structure,
as it will on railway over-bridges be a carbonaceous deposit
electro-negative to the iron, causing galvanic action and
therefore aiding corrosion, attracting moisture and all other
corrosive influences. A similar deposit will occur in other
smoky situations, as particles of soot or carbon rest upon an
iron or steel surface, sulphurous acid being present, and
moisture of the atmosphere; and any such covering will
almost always have a corrosive tendency.

In order to prevent galvanic action in the same plate or structure, it is advisable to avoid any variation in the electro-chemical properties of the metal. As wrought iron, steel, and cast iron do not possess the same electro-chemical properties, it follows that galvanic action, varying in intensity between any two, may be set up by mediate or direct contact. Thus, if steel is joined to wrought iron, the decomposing effect of galvanic action is in progress, and the iron may suffer; or steel or wrought iron to cast iron, but especially in the case of steel, as then by experiment the effect is increased, and is stated to be as violent as when copper is made an element in the galvanic circuit in connection with wrought iron. In brief, the direct, indirect, or circuitous contact of all metals of different conducting powers should be avoided as much as possible, or a durable insulator be placed between them, for paint is but a kind of expedient or remedy liable to be rendered useless by decay, and if there happens to be a bare spot it will be quickly corroded, such action will be concentrated on the exposed place, and a bare surface may easily be caused by abrasion, or during the course of erecting a structure. The following electro-chemical scale,* arranged after Berzelius, according to their relative degrees of positive and negative, and negative electro-chemical character, is useful for reference. It commences with those substances possessing the strongest electro-positive properties, and ends with those of the strongest electro-negative properties. Although several substances appear in the scale that are seldom employed in engineering structures, they cannot be eliminated without disturbing the scale. Some consider sulphur and nitrogen to be less negative than either fluorine or chlorine, and that hydrogen should be towards the end of the positive division, but the general comparative value of the scale is acknowledged. It has, however, been pointed out that the division between gold and hydrogen is in great measure an arbitrary one, and the use of the table is in indicating the general electro-chemical character of the substances.

* Vide ' Circle of the Sciences.' ¦

Positive End.

———

POTASSIUM.
SODIUM.
LITHIUM.
BARIUM.
STRONTIUM.
CALCIUM.
MAGNESIUM.
GLUCINUM.
YTTRIUM.
ALUMINIUM.
ZIRCONIUM.
THORIUM.
CADMIUM.
MANGANESE.
ZINC.
IRON.
NICKEL.
COBALT.
CERIUM.
LEAD.
TIN.
BISMUTH.
URANIUM.
COPPER.
SILVER.
MERCURY.
PALLADIUM.
RHODIUM.
PLATINUM.
IRIDIUM.
OSMIUM.
GOLD.

HYDROGEN.
SILICIUM.
TITANIUM.
TANTALUM.
TELLURIUM.
ANTIMONY.
CARBON.
BORON.
TUNGSTEN.
MOLYBDENUM.
VANADIUM.
CHROMIUM.
ARSENIC.
PHOSPHORUS.
SELENIUM.
IODINE.
BROMINE.
CHLORINE.
FLUORINE.
NITROGEN.
SULPHUR.
OXYGEN.

———

Negative End.

For instance, sulphur and chlorine, two of the most negative of substances, must be viewed as positive in relation to oxygen. In fact, each substance throughout the scale may be viewed as both positive and negative: positive in relation to those below it, and negative in relation to those above it; those of the upper end being strongly positive and feebly negative, and those of the lower end strongly negative and weakly positive.

The electric relations of metals, of course, vary if placed in certain solutions or mixtures, none of which are likely to be met with in ordinary engineering structures. The thermic condition has an influence in the electric conductivity of metals, as their order is not the same at all temperatures. At 60° F. the order for voltaic electricity, beginning with those which conduct most freely, is: silver, copper, gold, cadmium, zinc, brass, tin, palladium, iron, steel, lead, platinum, German silver, antimony, mercury, bismuth, potassium. By these two scales a fair indication of the effects of galvanic action, so far as regards the promotion of corrosion, may be generally gathered.

An iron plate placed in sea water will corrode more quickly than a copper one, but if they are in contact the corrosion will be very greatly increased, owing to galvanic action, and the copper will appear to be able to resist corrosion. It will be noticed in the scale that iron is considerably higher in the positive end than copper. The accelerated decomposition of the iron is thus explained. In the case of iron or steel, if the water is able to corrode *one* of the metals, the oxidising medium is in active operation, and accelerated corrosion, the result of galvanic action, is proceeding. Therefore an insulator is required to prevent the communication of electricity, but the importance of a *complete* insulating covering cannot be over-valued, because, if a bare spot should occur in an otherwise coated plate, galvanic action may become concentrated on that alone, which will then become quickly corroded.

Experiments have shown that carbon gives to steel some power of retaining magnetism, and that the permanent magnetism of steel is affected by the amount of carbon which it contains, therefore the strength of the current will vary, but the depth to which magnetism penetrates in iron or steel is a contested point. It exerts an influence on the corrosion of iron or steel.

The decomposing effect, or accelerated corrosion, arising from galvanic action may be stated approximately to be according to the juxtaposition of metals, the effect being the greatest when they are in close contact. Every metal being electro-positive to its own oxide, when steel or wrought iron with the oxide scale on it is placed in an oxidising liquid, as salt or sea water, the medium is present to at once promote corrosion ; further, it may occur between the metal of a plate and the oxide scale, and this has been proved to have been caused by the oxide formed in the early stages of manufacture being crushed or pressed into the metal in the rolls, resulting in the metal having impurities in it inducing galvanic action. In fact, a plate with the oxide scale on, and one without, if placed under the condition of electric action, as in salt or sea water, will be an electric battery, and the plate with the oxide coating will produce the necessary decomposing electric condition. Carrying this further, it will be readily understood that even some alloys, composed of metals of varying electro-chemical character, may, in an oxidising medium, as salt or sea water, be in an

I

electrically decomposing condition, inducing internal action
with deleterious effect. Usually, alloys are harder than their
components. It will be noticed that carbon is placed sixteenth from the negative end, and iron sixteenth from the
positive end, and, therefore, carbon is decidedly electro-negative to iron; consequently accelerated corrosion by galvanic
action will be set up by direct or mediate contact. In order
to lessen galvanic action as much as possible in engineering
structures, the accelerated corrosion so arising should obviously be guarded against by providing that, as far as
practicable, different metals should not be placed in close
union or in close contact, and that the metal is well covered
with non-oxidising paint, cement, or an insulating coating,
for, by the junction of such metals as steel with wrought iron,
cast iron with steel or wrought iron, electrical action is
greater than with metals whose electro-chemical properties
more nearly approach, the main point being to attain as
much similarity as possible. Local galvanism has also to
be considered, for the complex composition of steels causes
an interchange of the electro-chemical positions, and "it
may be inferred that the crystalline formations of their
structure nearer the surface are at first more readily acted
upon by sea water, until inner and seemingly less vulnerable
portions have been reached." When there is a variation in
the condition of a mass of iron or steel, one having less
affinity for oxygen than the other, contact of the former
makes the latter become oxidised more rapidly. The
positive metal, being more rapidly oxidised and acted upon
than the negative metal, also militates against an equi-
librium being maintained, and aids the alternate variation
of the electric position. The reliable recent experiments of
Mr. T. Andrews, F.R.S., M. Inst. C.E., of the Wortley Iron-
works, extending over 300 days, on steels and irons in
galvanic action immersed in clear sea water, have shown
that a complete interchange of electro-chemical position
occurs after very considerable intervals, affecting the rates of
corrosion, and that it is doubtful whether a permanent

position of rest finally ensues between the metals, and the probabilities are against it, though eventually the galvanic action becomes very small from the accumulation and possible influences of the oxides. He stated in a paper communicated to the Institution of Civil Engineers, that the conditions attending galvanic action between wrought iron, cast metals, and various steels, in sea water, is important, as the accurate measurement of the electro-chemical positions of the metals when in combination should afford an indication of their relative corrosibility under such circumstances in sea water; also, in a series of most carefully conducted and reliable experiments,* he "arrived at the following conclusions, namely: that wrought iron and various steels, when exposed singly and separately, without liability to galvanic action other than local, to the action of sea water for long periods, showed a greater corrosion on the part of all the steels than the wrought iron, the advantage in favour of the wrought iron as compared with the steels amounting roughly to 25 per cent. and upwards. It was also noticed that corrosion was increased in the steels in proportion as the percentage of combined carbon was greater. Further, it has been found that the galvanic action between wrought iron and steels induced a largely increased total corrosion in the several metals." Also, "that the corrosion of metals is considerably affected by stress, varying according to the nature and extent of the strain applied. It might have been thought that metals under stress would be more liable to increased corrosion than when in their normal state. His recent experiments, however, indicate the opposite conclusion. That is, when 'strained' is considered separately from 'unstrained' metal. When, however, the strained metal is in galvanic circuit or combination with the unstrained metal in any solution, an increased total corrosion ensues from the galvanic action which this research has shown to arise consequent upon the difference of potential between the two. The reason is that stress, whether tensile,

* 'Minutes of Proceedings,' Inst. C.E., vol. cxviii.

I 2

flexional, torsional, or of any other kind, considerably alters the physical properties of iron and steel. Stress increases the rigidity of both iron and steel and renders the metal harder, also greatly reducing its properties of elongation or ductility. A higher tonnage is required to break a 'strained' than an 'unstrained' portion of the same metal. A tensile stress applied to a wrought-iron shaft, producing an elongation of only 2 per cent., increased the tensile resistance of the metal 2·66 per cent. Other investigators have also noticed a similar alteration in the properties of metals referable to stress. From these observations it is manifest that the stresses applied to the metals examined for corrosion, altered their structure, render them harder in nature, i.e., the metals showed a higher tensile strength subsequent to the strain than previously, and hence less liable in the strained condition to be acted upon by sea water, or other waters, than in their normal or softer condition. The experiments, however, indicate that an increased total corrosion, in excess of the normal corrosibility of the metal, occurs in a metallic bridge, vessel, boiler, or other structure, from the action of the local galvanic currents which are hereby shown to be induced between 'strained' and 'unstrained' portions of even the same piece of iron or steel forging, bar, or plate. Hence a strain occurring in a metallic structure tends, owing to local galvanic action thus set up, to increase any corrosive forces which may be deteriorating the metal of which it is composed."

Thus to the fatigue of metals by stress may have to be added that of galvanic action aiding corrosion set up between the strained and unstrained parts of a metallic structure, especially when the strain is locally severe or of an uneven character.

Mr. Andrews's further valuable experiments to ascertain whether the normal corrosibility of metals might be considerably affected under the influence of strain, are novel and useful. They were made on small round bars. Plate $\frac{1}{32}$ inch in thickness. All metals were perfectly bright. As regards *tensile stress*, it appeared the *un*strained metal was being more rapidly corroded than the strained metal, which latter had in most cases been strained so as to produce an elongation of 20 per cent. between gauge points 3 inches apart.   In the case of *torsional stress* similar results occurred, viz. the *un*strained metal was being corroded the more. When the stress was flexional, similar results occurred.   In all cases the *un*strained steels were more electro-positive than the strained steels, and, therefore, were more acted upon by sea water.   Mr. Andrews states that "the electric measurements ought perhaps to be regarded as tentative indications establishing a general principle."

In the Journal of the Franklin Institute, Prof. Munroe, U.S.N.A., states that two steel chisels found in the U.S.S. *Triana*, were deeply corroded by the action of the salt water, but the corrosion was confined entirely to the soft metal, the tempered points not being attacked in the least; the corrosion being deepest at the line of contact between the tempered and untempered metal.   Whether the change which takes place in the tempering of steel is a chemical or a physical one it is not attempted to be shown, but it is considered the tempered steel is not so readily acted upon by salt water as *un*tempered steel, and it is suggested that when the untempered and tempered steels are brought into contact in the presence of salt water, they constitute an electro-chemical couple, and that this hastens the destruction of the *un*tempered metal.

Among some instances of galvanic action accelerating corrosion, additional to the familiar examples of iron and copper in contact in sea water, and the base of iron railings when fixed with lead run in a hole in stone, instead of by Portland cement, or in glycerine and litharge stirred to a paste, which hardens quickly, and will cement iron in stone ; or in sulphur, which has been recommended to be preferred to lead for preserving iron bars or staples that have to be fixed in stonework ; may be mentioned that Mr. Hyslop stated at a meeting of the Institution of Engineers and

Shipbuilders in Scotland, that he had connected the boilers of a steamer with a plate of zinc immersed in the sea, and, by the use of a galvanometer, he was able to observe the effect galvanic action produced at every lurch which the vessel received from the motion of the waves. Mr. Gilchrist, then President, said that galvanic action was set up in vessels having different metals used in the construction of their parts, was a very patent fact; for if a cast-iron screw were used on board a wooden vessel coated with copper it would, disappear in twelve months, falling to the bottom of the sea. Such vessels have brass instead of cast-iron screws. It has also been recorded that the replacing of the iron screw blades of a steamer by others of manganese-bronze, caused such corrosion of the steel and iron in the stern posts and surrounding parts, during one voyage between London and the Cape of Good Hope, that they had to be entirely renewed, and had now to be carefully protected by zinc. But, perhaps, one of the most striking instances of galvanic action is that related by Mr. John Donaldson many years ago. He then said the practice of his firm with steel plates had been to preserve, as far as possible, the oxide on the plates, with a view to prevent them from getting rusted, the oxide itself being considered to protect the plate covered by it from rust. Favourable accounts were at first received of a boat, the plates being as described, but one day she made so much water as to be in danger of sinking; the boat was then examined, and it was found that the plates in the bow and in the stern, and others in the boiler and engine room, were pitted with very small holes. Some of the black oxide had been knocked off in the process of working the plates, and the oxide left on had contributed to increase the rusting of other parts. It was the black oxide that had been acting *galvanically* in oxidising those parts of the plates that were exposed. When the plates of a boat were well painted and kept clean, very little of such action took place. Their practice after that experience was not only to remove the scale from the plates, but to galvanise the whole of the hulls, and since that had been done there had been no trouble as far as oxidation was concerned. Another instance of galvanic action, for there was no other way of accounting for it, that came before the author many years ago, was that of a small *steel* plated vessel which was riveted with *iron* rivets, the result being that the steel plate around the rivet-heads became sufficiently corroded to cause the heads and a small portion of the iron rivets to project beyond the surface of the corroded steel plates. The grooving of propeller shafts is considered to be generally due to galvanic action between the steel shaft and the sleeves forming the shaft journals in the stern tube. This has been proved to be the case, as the galvanic action has been prevented when the shaft has been covered with india-rubber or with a few turns of marline near the brass sleeves. Professor Lewis has demonstrated by experiment that of two steel wires dipping in sea water, the chemical action on one, which was kept in a current of hot moist air, caused it to be strongly electro-positive to the other. It is considered that similar chemical action would sometimes occur in hot moist steam. It has been found that while wrought-iron pipes taking condensed water from iron steam pipes connected with the same boilers showed no signs of rapid corrosion, those conveying the same water from a copper steam pipe were corroded through, and required to be renewed in a short time. This was probably due to galvanic action kept up by the constant circulation of steam and feed water between the two. Some cast iron coming from the blast furnaces of the South Oural Mountains in Russia was, on analysis, found to have the following composition :—

	Per cent.
Iron	83·5
Copper	8·2
Tin	1·3
Cobalt	0·5
Silicium .. `	1·0
Tungsten	0·1
Carbon	3·0
Manganese	2·4
	100·0

Under the microscope, on the metal being fresh cut, small grains of copper were easily remarked in the mass of the metal.* In such a metal galvanic action is probable, and it is to be regarded as one liable to accelerated corrosion, because the electro-chemical properties of iron, copper, tin are different.

Electric railway return currents have been observed to cause corrosion of adjoining water and gas pipes. A paper was read by Professor Jackson † on the subject, in which it appears that experiments were made at Boston, U.S.A., by connecting the reinforcing wire laid between the tracks to the water pipes, but the supplementary wire was destroyed in several places; then the polarity of the generators was reversed, the current being sent out through the rails, and back through the overhead trolley wire, but this current sent through the rails took to the water pipes and the lead cable coverings with disastrous results, following the law of divided currents, and leaving them at many points along the line, causing serious corrosion at these places. On the direction of the current being reversed, so great a current flowed along the water pipes that, at a joint where oakum was used for caulking, it was sufficient to set fire to the oakum. The loss of pressure on the return circuit was from 25 to 100 volts, or from 5 to 20 per cent. of the total pressure. On the water pipes being experimentally connected with the negative pole of the dynamo, a new danger occurred, for the difference of potential between the gas pipes and the water pipes caused a marked electrolytic effect on the former. By connecting the gas and water pipes together in all parts of the city, to arrest the action, fair results were obtained, but the expense to the city and to the company was great, and finally far from satisfactory to either party.

In Brooklyn, where there are many electric railways, numerous cases of serious corrosion had occurred, and the

* See ' Iron,' Sept. 2, 1876.
† See 'Journal of the Association of Engineering Societies,' Phila-delphia, 1894.

Board of Commissioners of Electric Subways of Brooklyn reported, in 1893, "that all kinds of buried pipes are being eaten away in many places. As an example, an iron service pipe, buried at a depth of 4 feet below the track, had been completely perforated in a month. At Milwaukee, the electrical engineer to the Street Railway Company reported, in December 1893, that at 200 feet from the power-house, a 6-inch water-main was so badly corroded, after the electric railway had been at work four years, as to render it entirely useless, and when taken out of the ground it was so soft in some places that a cane could easily be poked through it. The corrosion was arrested at Milwaukee by making numerous low-pressure connections between the pipes and the rails, thus keeping the two at the same potential, and by connecting both pipes and rails at the power-station to the negative pole of the generator. As much as 28 per cent. of the total output is now found to be returned by means of the pipes, and this plan has been working satisfactorily for more than a year. At Chicago, Professor Barrett reported, that in some experimental work, in which a current of 0·3 ampere continued for three weeks, it was most destructive to a lead telephone cable, while another, buried in the same soil, which was not subject to the action of the current, was unaffected. In Janesville, Ohio, a 4-inch cast-iron water pipe was completely perforated in two years. At Columbus, Ohio; Hamilton, Ontario; Indianapolis; Philadelphia; Los Angeles, California; and many other cities, where considerable electric railway systems are in operation, in every case the corrosion has exhibited the same general features. The iron pipes are usually "pitted" in many places.

It is now practically agreed that the reason for the extraordinary corrosion is the imperfect character of the return circuit of electric railways. As first constructed, the rails in connection with the surrounding earth were relied upon to carry all the current back to the generator. It was soon found, however, that the current would not confine itself to this path, and that the resistance of the earth was far from being as low as was originally supposed. Bending the rails, cross-bending, supplementary wires, and ground plates, were then tried, but have not answered, and the tendency now is to make the return circuit of fully as great conductivity as that of the overhead supply circuit, without relying upon any conductivity from the ground. There is little doubt that with a perfect return system, properly connected to systems of underground pipes, electrolytic disturbances will practically disappear in nearly all cities.

Two theories were put forward to account for the corrosion : (1) that it is simply due to chemical action caused by ammonia, saltpetre, leakage from gas mains, &c., found in the earth ; and (2) that it is the result of electrolytic action. Where chemical action alone occurs, water and gas pipes ordinarily last about twenty years or more, but this corrosive action has destroyed new pipes in intervals of from a few weeks to six years, and it has taken place where the electric current left the pipes. This was conclusive proof of electrolytic action, and the only question is, whether it occurs, (1) by direct electrolysis of iron, or (2) by the electrolysis of chemical compounds which are held in the water of the soil, causing secondary chemical reactions at the electrodes. A series of laboratory experiments, in which the practical conditions were produced as fully as possible, showed that the latter was the true solution of the problem, the iron pipe acting as the positive plate of a cell ; the water of the soil, with chemical compounds in solution, as the electrolytex, and the rail as the kathode, or negative pole.

Analyses of street soils usually show the presence of some soluble salts of ammonia, potash and soda. Experiments were made with such salt solutions, the result being that it was seen only such measures as will prevent electrolytic action of salts in solution in the soil can be relied upon to stop the corrosion of the iron pipes. Professor Jackson concluded by stating the solution of the matter is the proper construction and arrangement of the return currents. Alternating currents produce no appreciable electrolysis, and their employment would avoid all difficulty, but their use for driving street railway motors is not yet an assured success. Connecting the pipes with the rails by means of heavy cables at points where the former are positive to the latter, has hitherto proved the most effective method of prevention. The conductivity of the track circuit must be properly reinforced by feeders, so that an undue drop may not be experienced in the return conductors, and these track feeders should always be insulated and put on the line exactly as are overhead feeders, in order to save them from corrosion. In Milwaukee, with 125 miles of track, and over 200 cars in daily operation, at a cost of about 1600*l.*, such a connection of pipes and rails has apparently succeeded perfectly. Investigations have shown that when the negative pole of the generator is connected with the trolley, the pipes are positive to the rails over an extended outlying district, and corrosion proceeds over a large area ; while with the reversed arrangement, the dangerous area is concentrated about the power-station. The latter method of connection allows the difficulty to be most easily dealt with by making frequent connections of pipes and rails within the affected area.

CHAPTER IX.

THE INFLUENCE OF THE SCALE ON CAST IRON, WROUGHT IRON, AND STEEL, AS REGARDS CORROSION.

As there is scale or skin on cast iron, wrought iron and steel when manufactured, the question arises whether it should be removed, or be regarded as a natural preservative. Extensive use seems to show that the natural covering of cast iron as it leaves the moulds should be preserved, although the appearance is not so good ; and that cast iron is more durable if the skin is allowed to remain than if the metal is planed or manipulated, as it is generally found to be a film of silicate of the protoxide of iron, and is a protection. As silica is not quickly acted upon by salt water, in all cast-iron work that will be submerged and cannot be periodically inspected, the preservation of the natural skin of cast iron which it obtains when run into a sand mould appears to be advantageous, particularly if it is to be immersed in salt water, and it is a protection which, with care, can be preserved ; but silica is only an apparently insoluble substance in water, for Mr. Crookes, F.R.S., and Professor Odling found, in experimenting for the Huddersfield Corporation in 1882–4, that silica dissolved in water to an appreciable extent, from $\frac{1}{10}$ to $\frac{7}{10}$ of a grain being found in a gallon of water.

It is important that the silicate scale on cast iron, which is harder than the metal it covers, being fused sand or loam on the molten metal, should be coated with oil or paint soon after casting, and before any oxidation has occurred ; as unless it is protected the metal underneath it will become exposed as the silicious film may fall away. If oxidation of the fused skin be allowed, unless the rust is carefully removed, paint spread on its surface will in time become detached. When any portion of the scale of cast iron is removed by manipulation, turning or planing, vulnerable places for active corrosion are produced.

The skin on wrought iron is a chemical combination of iron with oxygen, the proportion of the latter increasing as the iron is exposed to the air. It will fall away sooner or later. With regard to the scale on wrought iron or steel, differences in the process of manufacture and in the nature of the metal cause it to be necessary to remove it in order to effectually prevent corrosion, or paint the metal, and it is decidedly injurious to wrought iron or steel. Iron and steel not being manufactured in the same way, their surfaces are not alike. Steel cast in an iron ingot mould, and cast iron in a sand or loam mould, have dissimilar substances in contact with them during manufacture, and steel which is reheated and passed through a rolling mill has a different kind of scale to wrought iron. Any attempt to preserve the entire skin of wrought iron as it leaves the rolling mill can only end in leaving some portions covered and others bare, cause accelerated and concentrated corrosion in the bare places, and result in ultimate failure, for as iron in passing through the rolls is exposed to the air, oxidation really commences, in however small a degree, in passing the rolls, and any such film of oxide is unstable, and may flake and become detached, whether it is covered with paint or not. In the case of wrought iron, the scale will fall off at some time or other, depending upon the magnitude of the rusting that took place during the process of manufacture and upon the homogeneousness and general quality of the metal. If a complete and equable covering could be so obtained and maintained it would form a protection, and most probably an air-tight coating. In practice, however, corrosion will continue with any such scale, and the process of oxidation will rather be accelerated than retarded by it. In fact, quite distinct from their chemical composition, the skins

of cast iron, wrought iron, or steel, are dissimilar, because
of the difference in the process of manufacture, and the
latter requires to be considered, no matter what method of
production may be adopted.

That galvanic action is set up, when a rolled plate is
immersed in sea water, between the black oxide scale and
the iron, may be considered as agreed ; but, if a bare plate
were immersed in sea water, i.e., one with the oxide scale
removed, and one with it remaining, the oxide scale would
form a kind of *temporary* shield. However, the black oxide
scale is *not permanent*, and therefore although, for a short
time, it *may* help to protect the surface, when air permeates
to the iron, the scale gradually flakes or falls off with any
paint that may have been spread upon it. If the surface of
the iron, when so bared, was properly prepared by rubbing
or other approved process, it would be preserved, *but only
when the black oxide scale had been entirely removed*, consequently
it is far better to remove the scale *at first*, before painting,
than to wait till it falls away with the paint, which would
be when a structure is in use, and then, probably, there will
be places almost inaccessible to painting and thorough
inspection, and the work cannot be done so easily or so
effectually.

The ordinary cheap market wrought iron has a thicker
scale than the higher brands, and may have laminations
and perhaps cinder in it, but the cinder in iron is generally
regarded as dielectric ; however, if it be a carbonaceous
deposit it may act deleteriously in another way, namely, by
becoming a nucleus in which moisture and acid may be
retained and become condensed. In the case of wrought
iron, careful consideration confirms the view that it is the
best plan to remove any thin scale that may be upon the
surface, for it cannot be a preservative ; it in no way adds to
the strength, and, moreover, is a cause of accelerated corrosion.
This requirement is no mere mechanical refinement, and,
in course of time, may become the universal practice, for
metallic structures generally have hardly been introduced

sufficiently long to cause marked attention to be given to corrosion in its various aspects, but every day must increase it; and as iron structures, especially bridges, have to be removed or altered to comply with increasing traffic, ocular demonstration may prove that more attention and special precautions to prevent corrosion may be necessary. With respect to the skin or scale on steel plates, experience shows that the hard black oxide scale on steel is very deleterious in its effects, and more seriously so than in wrought iron. In fact, it approaches that of copper, and on high authority it has been stated, " a tolerably compact coating of oxide is as detrimental to steel exposed with the oxide on as copper." Dr. Siemens said it was of much importance to perfectly clean the surfaces of plates, for if they were carefully cleaned of oxide by dipping in the first instance in an acid solution, it was found corrosion was always much lessened, the reason being that the scale on steel which was produced in rolling, being a magnetic oxide, negative to the steel, had a very deteriorating influence on it; or if any scale were rolled into steel plates, as occasionally occurred, when the metal was exposed in the presence of such magnetic oxide, corrosion rapidly followed. By removing this skin in steel, this cause of rapid corrosion is prevented. The Admiralty require the magnetic oxide scale to be removed from steel plates, for it was found that unless it was completely cleared, rust formed under the paint when carefully spread on such scale, and also that unless the surfaces of steel plates were carefully cleared of the black oxide produced in the rolls, corrosion was not uniform, but when the scale was removed corrosion was very uniform in salt water. If the surface oxide is *not* removed, the effect of the oxide in the neighbouring bared metal is as strong and continuous as copper can be. In some locomotive works, the scale of all steel plates is removed either by bathing them in or brushing them over with a solution of sal-ammoniac. It is not only found that this much reduces the tendency to corrosion in the boilers, but that more secure and water-tight joints can then be made.

This is a decided advantage, and applies equally to girder-work and work made up of plates, for the closer the plates are together, and the tighter the joints and edges, so will active corrosive influences be hindered from coming into action.

It has been asserted that if the magnetic oxide scale on a steel plate could be preserved and made to adhere to the solid metal it would be and is a protection against corrosion, but in practice, either from blows or in putting the parts together, patches fall off, even if they do not when untouched, and leave the paint bare, and even do so when the scale has been painted, and so centres of corrosion are formed. For these reasons alone, and apart from the question whether oxide scale may be a protection or not under certain circumstances, the better and safer plan is to remove it by pickling or brushing; if not, blows, vibration, or attrition may do so and expose a plate.

How can the scale or the film of oxide on wrought iron or steel be removed? By the metal prepared for coating being dipped in a bath of dilute acid, or by a wash of weak acid, and afterwards by careful neutralisation, washing and drying; but this process is not so complete and certain as planing, filing or grinding, the latter two methods being almost confined to test pieces. On account of expense, pickling the plates, and so removing the oxide, is the way in which, in ordinary engineering structures, the thin skin can be removed at trifling cost. Some experiments made by the Dutch Government with respect to the preservation of plate iron in railway bridges gave the following results. Sixteen wrought-iron plates were cleaned by immersion for twenty-four hours in hydrochloric acid, they were neutralised with milk of lime, washed with hot water, and, while warm, dried, and washed with oil, and coated as described. Sixteen were merely cleaned mechanically, by scratching and brushing; four sheets of each kind were then exposed to the weather, and examined after three years. The following were the results of these experiments: (1) The red lead was perfect on both sides, so that it could not be said whether the chemical cleaning was any use. (2) One kind of iron oxide red paint gave better results on the chemically-prepared plates than on the others; but another kind of iron-oxide red paint gave not very good results on the plates only scratched and brushed. (3) The coal-tar was considerably worse than the paint, and had even entirely disappeared from the iron sheets which had not been treated chemically, but only been cleaned by brushing.

In Holland, especially, great care is taken to remove the scale before painting. Some Prussian experiments were recently made on the effect of pickling and rusting on the strength of iron; and on steel rails of the State line section; wrought-iron joists, $9\frac{1}{2}$ inches by $3\frac{1}{2}$ inches; wrought-iron round bars, 0·8 inch diameter; and steel round bars, 0·8 inch diameter. The pickling was done with sulphuric acid, diluted with water in the proportion of 1 to 100. The metal was protected from direct attack by covering one end with melted zinc. The time of immersion was seventeen hours. After pickling, some pieces were immersed in lime water for many hours, so as to remove all traces of acid. The results were considered to show that, as regards chemical composition, the susceptibility of iron to become brittle by pickling and rusting is least in cast iron and silicon steel, and highest in wrought iron, and, according to Bädecker, in high-carbon steel. Combined carbon appears to increase the action, and silicon to diminish it. The pieces were treated in the condition as delivered, after exposure for the purpose of rusting; galvanised and tested at once, galvanised and exposed for a time, pickled in acid and immediately tested, and pickled and kept for a time in a dry place. It was thought, from some results of the experiments, that acid sets up chemical changes in wrought iron which it does not in steel; however, the steel treated was exceptionally rich in silicon, no doubt adding to its resisting power. These experiments are referred to to record them here, and not to

K

consider them as conclusive, nor are they so claimed, although they are well worth attention.*

The effect of pickling iron wire in water acidulated with sulphuric acid, in order to remove scale, has been shown by Prof. Hughes and others to be to make it brittle.† It did not affect its tensile strength, *so long as the metal was not sensibly corroded*, but the extension under stress and the capacity of resisting bending strains were notably diminished; and an action similar to that of weak acid is produced by the atmosphere when iron and moderately soft steel are exposed to it in an unprotected condition. Contact of the iron with zinc, which renders the former electro-negative, increases the influence of the acid. Brittleness produced by pickling or rusting is removed by annealing, and also disappears, or is considerably diminished, by allowing the brittle metal to rest for some time in a dry place. Cast iron is not sensibly, or only very slightly, affected by pickling. The pickling was in 1 per cent. sulphuric acid water for twenty-four hours without previous cleaning; tests after three days. By pickling in 2½ per cent. acid for twenty-three hours after cleaning the surfaces with ether. The mechanical tests were applied immediately. These experiments show the important effect of sulphuric acid on iron and steel.

Pickling in acid, then carefully neutralising it, washing and drying, may be considered as the most practical, sufficiently trustworthy, effective, and cheap process yet discovered of removing the scale.

---

* *Vide* 'Mittheilungen aus den Koeniglichen technischen Versuchsanstalten zu Berlin, 1890, Supplement I.
† 'Stahl und Eisen,' vol. vii.

131

CHAPTER X.

THE SERVICEABLE LIFE OF METALLIC STRUCTURES, AND SOME
EXAMPLES OF CORROSION.

THE serviceable life of iron or steel *cannot* be determined,
although, by deduction, it may be approximately estimated.
It would be useless and misleading to attempt to establish
any rule, the circumstances and conditions being so dissimilar.
The corrosive influences to which each metallic structure is
subject, and also the other conditions, will affect the decay
and durability of the metal.

Metallic bridges are of recent origin. The first erected
in England was Pritchard's celebrated cast-iron arched-ribbed
Coalbrook Dale Bridge, over the river Severn, constructed
about 116 years ago, which was followed by Burdon's cast-
iron arched-ribbed bridge over the Wear, at Sunderland,
some ten years later. Similar bridges were then built by
Telford, over the Severn, at Buildwas, Shropshire ; by Rennie,
over the Witham, at Boston, Lincolnshire ; and, in 1815–19,
Sir John Rennie erected his masterpiece, the Southwark cast-
iron ribbed bridge over the Thames, at London. The first
boat made of malleable iron plates is generally acknowledged
to be that constructed by Mr. Manby in 1820–21, although
at Broseley, Shropshire, John Wilkinson, ironmaster, is said
to have launched the first iron boat, but the system of construction is not clearly stated. Several vessels were subsequently built of iron plates, by Messrs. Maudslay and Field,
about the years 1830–31, and by 1846 about one hundred
such British-owned vessels were afloat. Compound bridges,
i.e., those constructed of cast-iron girders and wrought-iron

K 2

bars to truss them, received their doom in the failure of a bridge over the river Dee in 1847. With regard to wrought-iron girders: the first employment of wrought iron in the present form of plates and angle-irons was by the Messrs. Fairbairn, about the year 1832, and, with the exception of a few suspension bridges, malleable iron bridges were hardly known in 1837. It was not till Sir Wm. Fairbairn, in 1846, erected for the late Mr. Vignoles, Past-President Inst. C.E., the wrought-iron tubular girder bridge over the Liverpool and Leeds Canal, for the Blackburn and Bolton Railway, that wrought iron may be said to have commenced to be practically applied on a considerable scale to such important engineering structures as girders. Its practical application in the form of plates riveted to angle or T-irons, &c., may now be said to be about sixty-three years old.

We thus have evidence that, with due care, *un*submerged and properly coated cast iron in arched bridges may be said to be practically not deteriorated from visible corrosion after 116 years, and wrought-iron plate bridges after about fifty years, taking the first very large wrought-iron plate bridge, that of Robert Stephenson's Britannia Bridge, erected in 1847–50. The application of *steel* plates to girders, ships and important structures, is quite of recent origin. Cast-iron sheet piles, whether in cofferdams or wharf walls, may be said to have been used since about 1815, and largely between the years 1820 and 1834; and solid, hollow, and screw, cast and wrought-iron piles extensively during about the last fifty years. With the exception of a few small pieces of ironwork embedded in mortar, occasionally found in some ancient building or ruin, or in a wall, or an old gate here and there, very little direct and unquestionable proof exists absolutely testifying that any ironwork used in the form adopted in modern engineering structures, and similarly exposed to corrosive influences, will last 200 or 300 years or more.

In order to show how the serviceable life of metal in a form largely used can be prolonged, it is but necessary to mention that, during the construction of the Britannia

Bridge over the Menai Straits, some rejected plates were left entirely unprotected upon a wooden platform exposed to the wash and spray of the sea. In two years after, they had become so corroded that they could be swept away with a broom. The plates were by no means thin, as they varied from $\frac{7}{16}$ to $\frac{3}{4}$ of an inch in thickness. The late Mr. W. Baker, when chief engineer of the London and North-Western Railway, found the corrosion which had taken place in the plates of the Britannia Bridge, exposed to the smoke and fumes of the locomotives, all being carefully painted and maintained, would give no less than 1200 years as the time before the plates would be entirely corroded ; this would be $\frac{1200}{2} =$ 600 times more than that of the rejected bared plates left on a stage on the shore as described. Further, with proper precautions Mr. Baker and Mr. Ramsbottom "could not give *any* practical limit to the endurance of the Britannia and Conway tubes." These facts are here mentioned to show that in very great measure the serviceable life of a metallic structure, so far as corrosion is concerned, can be prolonged to almost any practical limit by constant care of the surfaces, and repainting with an anti-corrosive coating, containing nothing that will produce or induce any deleterious action in or upon the metal.

If bridges, roofs, promenade and landing piers, and similar structures, had the care bestowed upon them that iron and steel ships receive, their serviceable life would be much prolonged.

It cannot be too much borne in mind that corrosion is very much accelerated by allowing it free action for a time ; in fact, reliable experiments have shown that the corrosion of plates of cast iron, wrought iron and steel, especially the steel, rapidly increases with time ; for it has been found in some experiments that the second year's corrosion compared with the first year's was 50 per cent. greater. This accelerated corrosion can be almost entirely prevented by reasonable care and attention, and at little expense. It is, therefore, well to remember that, before the introduction of plates of various forms for the floors of bridges, the necessity of renewing the floor planking or timber platform at intervals caused more frequent uncovering of the main and cross girders than is now requisite. Bridges of large span may be said to be constantly watched and maintained, and it is not to them especial reference is here made. The serviceable life of iron and steel bridges, roofs, columnar or pile structures, may be much reduced by inattention to their preservation from corrosion, complication of the parts, badly-fitting rivets, rough butt-joints, the main girders being of loosely built-up plates; also from members of a structure being of such shape, or so, when fitted together, as to be gatherers of moisture and dust, and receptacles for the promotion of decay and decomposition.

In bridges, roofs, and unsubmerged structures, the extent of appreciable corrosion chiefly depends upon the attention given to them, the removal of the oxide scale before painting, the complete covering of the metal with an anti-corrosive and insulating coating, preferably possessing the power of hardening under water, the quality of the material, its freedom from galvanic action, and the nature of the atmosphere and climate in which the structures are erected. In fact, their serviceable life is governed by the care bestowed upon them in keeping them clean, and entirely and equally coated, which necessitates frequent examination, and renewal of the protection of the surfaces. Then, and then only, may their serviceable life, so far as regards corrosion, be considered as long-continued. In the case of submerged structures, the circumstances are different, for, except by divers, or by the use of compressed-air apparatus in some form or other, they cannot be inspected, nor can they be repainted in water, and the subterranean portion cannot be examined. The duration of the parts of a structure which are either constantly submerged or buried in the earth, as in the case of piles or columns, can only be deductively estimated from the behaviour of similar works subject to like conditions and circumstances. An economical method of testing whether the metal in a pier has corroded to a serious extent below low water, is to suspend, when a structure is erected, a few short lengths of piles of the same dimensions, form, &c., under precisely similar circumstances to those of the permanent columns, taking especial care that no galvanic action is set up. These lengths can be secured to the permanent work by an insulating medium at the shore and sea ends, and middle of the pier, or where desired or thought advisable, and a few piles can also be screwed or otherwise inserted in the ground. They can then be removed at any time for testing or inspection, and be replaced without interfering with the permanent structure, and will indicate the probable condition of the submerged and buried piles.

In a paper read before the American Society of Civil Engineers, by Mr. John D. Van Buren, it was stated, "Submerged bolts and other wrought-iron parts are generally badly corroded in less than twenty years; the corrosion taking place in lines parallel with the fibres. Certain kinds of cast iron could, perhaps, be made to last fifty years; but already (1875) authentic rumours are afloat that our—i.e., American—cast-iron lighthouses are materially injured by corrosion; fifty years is a generous allowance, and probably greatly exceeds the average life of cast iron in salt water."

The cast-iron cannons of the *Royal George*, which went down in 1782, and the *Royal Edgar*, were found to be quite soft, and like plumbago in some cases. They had been immersed for 62 years and 133 years respectively. The original cast-iron sluice-gates of the Caledonian Canal also corroded seriously, and had to be replaced; on the other hand, the pier at Milton-on-Thames, erected in 1842–44, designed by Mr. Redman, was supported upon cast-iron columns 3 feet in diameter and 1½ inch in thickness; and Gravesend Town Pier, designed by Mr. Tierney Clark, erected about the same time, and 1¼ inch in thickness; and the Maplin Sands Lighthouse, also constructed about the same time, of ¾-inch metal only, appear to be *un*affected. Sir J. Brunlees remarked, some sixteen years after the Morecambe Bay Viaduct had been erected, that there was no sensible decay in the columns. The whole of the cast-iron work was protected with a preparation of tar and asphalt. The tar was heated in a large tank, and each piece of ironwork was boiled in it for half an hour, as he believed it was the best preparation for coating cast iron. An examination was made by the chief engineer to the Bombay, Baroda and Central India Railway Company, of the condition of the piles of the South Bassein Bridge,* some of which had been fixed 25 years, and exposed during this period to the action of sea water. Specimens were cut from each pile that was considered likely to be corroded, and from an examination of these it was concluded the greatest corrosion in cast-iron piles exists close to low water, and does not extend to any considerable depth below it, a conclusion which also applies to bolts and braces. After 25 years' exposure in a salt-water way the piles were in very good condition, and corrosion has only occurred in places which are easily accessible for repairs and renewals.

The Austerlitz Bridge, Paris, erected in 1802–5, which consisted of five cast-iron segmental arches of 106 feet span, was taken down in 1854, and it was found that the cast-iron arches had not given satisfactory results, but were displaced and broken in places. The condition of some of the original cast-iron bridges taken down for the London and South Western widening works over Upper Kennington Lane, South-Lambeth Road, Wandsworth Road, &c., London, was not found to be satisfactory, nor was it expected it would be very reassuring.† It was stated it was caused by defects in the castings, such as air-holes, also that, "taking into consideration the condition and design of these ribs, and having carefully examined sixty after being broken up, it suggested that calculations as to their working strength must obviously be misleading, and that it is impossible to assign to them a proper margin of safety."

* Vide 'Engineering.' February 3, 1888.
† Vide 'Minutes of Proceedings,' Inst. C.E.   Paper by Mr. Szlumper, M. Inst. C.E., vol. cvii.

At Lowestoft, where cast-iron sheet piling, 1½ inch in thickness, after some forty years' immersion, was examined; it was found the iron had not become soft, and had only deteriorated slightly at the edges of the piles, where the original skin of the casting had been removed. An examination of the cast iron Fleet sewer, erected by the Metropolitan Railway Company, London, showed that the cast iron of the original crossing had suffered no deterioration from the constant flow of sewage matter through it; the red lead with which the interior had been cased was still adhering to the sides, although it had been washed away from the invert; all the joints which had their surfaces truly planed, and were made good with iron cement, were perfect, and the packing was as good as when it had been put in some four or five years before.

Professor Kennedy has mentioned that he had to inspect a piece of good wrought iron which had been erected 50 years before. He found that on its being broken across it showed signs of damp having penetrated it in many places, through defects in the welding of the original material; although it was not corroded on the surface, it being preserved by painting, particles of golden rust were found in the heart of the material. Moisture had found its way into the spaces in the metal where the layers of dirt, during the process of welding, rolling and manufacture, had originally interrupted the continuity of the metal; hence the more pure and homogeneous the metal, the less the chance of corrosion. This shows that, even although the surface of iron may appear to be sound and not corroded, it is possible for internal corrosion or electrolysis to be in progress.

It would be a valuable guide, when old girders are removed, if even a few bare details were supplied, stating the dimensions of the original members of a structure, and they were compared with the thicknesses when the bridge was taken down ; the nature of the traffic; abstract of specification ; date of erection ; how often painted, and with what substance, &c. &c. Then, in time, some valuable data could be tabulated showing the most vulnerable parts of an ordinary metallic structure, similarly circumstanced in every way, and to be erected for like purposes of traffic. At present, information is very fragmentary and difficult to obtain. Mr. Ewing Matheson, M. Inst. C.E., has mentioned that in altering a wrought-iron bridge at Stepney, erected 25 years before being taken down, and subject to a heavy and constant railway traffic, the upper boom or box had been riveted and caulked like a boiler, and was perfectly air-tight, and the inside plates were as good as on the day they were first put in; while some of the parts exposed to the atmosphere, and ineffectively painted, had deep pits bitten out in all directions, materially weakening them. The worst part was where the iron had been brought in contact with wood, the acid of which had so destroyed it that an angle-iron $\frac{1}{2}$ inch in thickness was worn down to a knife edge. In another case, an approach to a large terminus, the rivet heads were found to be almost entirely corroded, and the T-iron stiffeners were nearly rusted away. The following detailed description of the condition of the structure, erected about the year 1847, so far as regards corrosion, is extracted from 'The Engineer,' May 26, 1876, in which full particulars of the bowstring bridge, and a drawing were given.

"Generally, the ironwork of the main girders was in a very fair state of preservation, but it was much corroded wherever it had been in contact with wood, especially where the latter was decayed, as it was in almost all cases where it touched the iron. The decaying wood seems to have accelerated the oxidation of the iron, which had gone on to a greater extent where the two materials were in contact than can be accounted for by the simple retention of moisture; parts equally unfavourably situated in this respect being much less affected, as, for instance, where the iron was in contact with earth or rested on brickwork unprotected from the wet. These remarks, however, do not apply to portions of the top flanges of the cross girders. To form a bed for the corrugated floor-plates these flanges had pieces of plank wide enough to extend a little over both edges, bolted tightly down to them, and wherever these planks remained sound they had excluded the wet, and prevented the iron from being much rusted. There was a considerable amount of rust in all the confined spaces open to atmospheric influences, but not so much as in the former case; whilst inside the arched boom, which is completely closed, there appeared to have been no oxidation whatever, the inner surfaces of the plates remaining as clean as when they left the maker's yard. So entirely had this part escaped, that when the first section, $\frac{1}{3}$ of the entire boom, was lifted out, the threads of some screws which were seen projecting about half an inch through the plate into the closed chamber were still bright, having undergone no change except a very slight browning since they were put in to make the joint nearly thirty years ago. In some of the exposed parts of the main girders the rust had eaten out a number of curious little pits, of a more or less circular outline, varying from $\frac{1}{8}$ inch to 1 inch or more in diameter, and often fully $\frac{1}{8}$ inch deep; some of these pits were isolated, others were clustered together in considerable numbers; they were chiefly found in those parts of the main girders which were otherwise least injured by the rust. Some of these pits must have been of long standing, for they were painted over, and the paint had, in some cases, protected them from further action. The lower flanges of the cross girders were made of five thin plates, which were held together by rivets so far apart that the rust had got between the plates and thrust them asunder, so that in many places the part of the flange between two plates was bulged out so as to be much thicker than it originally was. The bulging out of the plates was especially noticeable at the ends of the girders, where, in consequence of the flange having been diminished in width, there was not room to continue the outside rows of rivets. This serious defect in these girders proves the necessity for having flanges which are to be exposed to the weather secured with a sufficient number of rivets to make tight the joints between the plates. The wind-bracing was quite ineffective, being only of flat bars 5 inches wide by $\frac{3}{8}$ inch thick. It had sagged in some places as much as 6 inches, and in one or two places was rusted entirely through.

" The condition of this bridge suggests the following observations on some points of practical detail, which may perhaps be advantageously considered in designing structures of iron combined with other materials. Ironwork, which is to be placed in situations exposed to the weather, should not be enclosed by wooden panelling, or otherwise covered with wood, where it can be avoided; and where it rests on masonry piers or abutments the iron should either be left quite open for the free circulation of air round it, or the masonry should be so built up to it as to exclude the air altogether ; confined spaces open to atmospheric influences should be avoided, not only as being inaccessible for repainting, but because they are attacked sooner than those parts which, being freely exposed, are soon dried after a wetting, and from which deleterious matter is periodically removed by wind and rain. Very thin plates are open to the objection that they present a number of joints which it is much more difficult to rivet tightly enough to exclude air and moisture, than in the case of thicker and, consequently, stiffer plates ; every joint is thus liable to be speedily attacked by the rust; and, for the same reason, when corrosion has once set in, they offer less resistance than thick plates to the increasing thickness of the layer of rust between them. This is well seen in the example before us, in which the upper flanges of the cross girders are each made of two $\frac{9}{16}$-inch plates, into the joints between which the rust had made but little way, whilst there was none of the bulging out of the plates which might be seen in many places along the lower flanges, although, in consequence of the top flanges having been covered with timber, they were much worse rusted than the bottom ones wherever the timber was decayed. The brightness of the screws inside the arched boom noticed above offers a sufficient testimony to the value of good work; in this instance the riveting had been so well done as to render the inside of the boom practically air and water-tight, so that it would probably have remained intact for an indefinite period. It is worth noticing that under the woodwork and in some of the confined spaces there was under the red rust a layer of imperfectly-oxidised iron, sometimes more than $\frac{1}{4}$ inch thick, which was very hard, and could not be removed by scraping, but which, on being struck sharply with a hammer, fell off in thick flakes; these were quite brittle, and had a rather metallic-looking fracture. By well hammering the iron it could be freed entirely from this scale, leaving a surface somewhat uneven, but quite clean, and with a greenish-black lustre. This suggests a point which should be looked to in repainting rusted ironwork, as of course any paint laid on this scale would speedily be destroyed. The ironwork of this bridge had been painted first with red lead, and afterwards with stone colour, which had protected it fairly well in those parts that were favourably situated, but a comparison of these portions with those so placed that the paint on them could not be renewed, shows how great is the damage which outdoor ironwork sustains if it is not kept thoroughly painted. The floor not having been made approximately water-tight, the cross girders had suffered greatly in consequence, the drip from the engines having kept them in some places almost constantly wet."

Mr. J. W. Wilson, in a paper read before the Society of Engineers, in March 1875, stated that the solid wrought-iron piles at West Brighton Pier, which was opened on October 6, 1866, were originally of a diameter of four or five inches. They had proved unsatisfactory owing to the continuous corrosion, as in several instances after a lapse of time, they were reduced to a diameter of two inches, or even less, and are being gradually withdrawn and replaced by a more durable material, cast iron. Examination of some iron railway bridges by Austrian engineers, consequent upon a State order in 1887, showed that in the case of wrought-iron box girders, open at the ends to the free circulation of air, the original colour has still remained after the lapse of 30 years, but it was found locomotive smoke seriously affected the girders under which the trains passed. The girders of the original Tay Bridge, erected some four or five years before, that remained after the disaster of December 28, 1879, which befel a portion of the structure, were found on inspection to be generally in good condition. Mr. St. John Day, in a paper read before the Institution of Engineers and Shipbuilders, Scotland, in February 1880, stated " that the bolts, $1\frac{1}{8}$ inch in diameter, holding the ties to the lugs of the piles of the original Tay Bridge were so corroded that they would have had to be replaced every four to six years, and that one of the bolts found was nearly, if not quite, half-way converted into rust." At the new Tay Bridge, the Board of Trade prohibited the use or continuance of small cast-iron columns with wrought-iron ties.

In a report of Colonel Macomb and Lieut.-Col. Ludlow, Corps of Engineers, U.S.A., on the Improvements of Rivers and Harbours in New Jersey, Pennsylvania and Delaware, 1879–80, the following paragraph occurs with respect to the appearance of some of the screw piles which were lifted and reset at greater lengths, to meet some contingencies of soft stratum, after an exposure of five years to the action of the salt mud and water. "Below the line of the bottom, the outer scale of the *wrought-iron* pile itself was still smooth, hard and bright, with only a few rusty spots here and there. From the level of the bottom up to low water, the surface of the pile—after the mussels, which covered it to the thickness of six inches with a closely-adhering coat, were scraped off—was found to be full of small cavities, and of a streaky fibrous appearance. Between low and high-water marks the pile was still smooth; but above high-water mark to the cap, corrosion had taken place, imparting a scaly blister-like appearance. The braces near to and above high-water mark had much corroded; the screw threads are cut perpendicular to the fibre of the iron, and nearly destroyed, making it necessary to cut them off, and weld on pieces for the new threads."

Inspection of suspension bridges erected 25 to 30 years ago has shown the imperative necessity of having *all* parts of a cable accessible for examination. Several wire-rope bridges have been rendered unsafe for traffic in less than 50 years, and had to be renewed owing to unperceived oxidation in the interior of the wire cables. The difficulty in preventing wire ropes deteriorating from corrosion is that in round wires there are spaces between them, and these must be completely filled with some substance of an elastic but tenacious nature, having great adherence—qualities somewhat contrary. It is advisable not to employ wire ropes in places that cannot be properly inspected. For this reason, in some modern wire-rope suspension bridges, links are used in the tunnels and wells, and anchorage abutment chambers, where corrosive influences will be very active owing to the humidity of the air, and moisture. Laminated-wire rope has been introduced, in order to remove the objection to round-wire rope under the circumstances mentioned, and to obtain compactness. Unless the interstices between the wire cables of suspension bridges are completely filled with an anti-corrosive composition, having the requirements stated, and the outer surfaces are properly protected, such cables will have their useful life soon impaired, and ultimately be so weakened as to be worthless. Galvanised-steel wire ropes are, therefore, sometimes used to prevent or lessen corrosion. There is, however, one point in favour of a cable bridge, namely, that the surface exposed to corrosive influences is very small as compared with, what may be called, rigid girders or arches of any design, and it is desirable to reduce the surface to be painted to a minimum; on the other hand, any corrosion is of *more* importance, and if the cables are neglected, and allowed to corrode, the whole stability of the bridge is soon impaired, and may be destroyed. Frequent inspection of the junction of cables with the anchorages of any wire-rope suspenders should be made; also at their junction with the stirrups and any woodwork.

The cutting action against one another of the wires forming ropes used for hauling or lifting purposes can be moderated by oiling. The oil should be such that it will permeate the rope. The principal causes of the destruction of wire ropes are the wearing of the outer surface of the outside wires, the rubbing of the wires against one another, the strain on the metal when worked over too small a pulley, and corrosion. It is necessary that the oil contains no acids, or corrosion of the wires will be accelerated. Wire ropes used on inclined planes do not generally wear out by abrasion, but by crystallisation. As steel-wire ropes, because of their lightness and great strength, are so frequently used, any reduction of their sectional area by corrosion is of much importance. It is especially necessary to keep them in a dry place, at a moderate temperature, and free from vapour or steam. It does not follow because a wire rope is well-greased that it is thereby preserved from corrosion, even if the grease is free from acid and of excellent quality, for if it hardens, rust may form between it and the wire. The mere covering a piece of metal with some substance which soon changes its character will not prevent corrosion.

In wire-rope suspension bridges, it is usually the case that those parts which cannot be inspected are most exposed to damp and deteriorating influences. In inspecting several such bridges, with a view to ascertain their condition, it has been noticed when a slimy liquid of a reddish colour appears on the surface, it is almost always an indication that the interior wires have been seriously attacked by rust, and it has been found to be so in several wire-rope bridges. M. Bernadeau, in the 'Annales des Ponts et Chaussées,' 1881, refers to a case in which only 15 out of 180 wires forming each cable were in good condition, the rest being as brittle as glass. The suspension bridges had been erected between 26 and 39 years. Two such bridges fell after 26 and 28 years' use. Repairs were effected by removing the rusted wires, and replacing them by new ones, the sound wires being carefully cleaned and dipped into boiling linseed oil, and finally the cables were coated with coal-tar. The operation was carried out without stopping any but the heavy traffic, only one vehicle being allowed on the bridge at a time. Wires have also been considered protected by dipping in asphaltum. It has been stated by an authority that there are some 500 suspension bridges in France. The fact that a less weight of wire is required than that of iron links, led to wire cables being very largely adopted instead of bar links, of which latter system the bridge over the Thames, at Hammersmith, is an excellent example. It has been found in France, and it may be also said in America, that the life of wire cables is precarious, for oxidation proceeds in the interior of the cable unperceived, and there is no proof that the means adopted for their preservation will do so after many years, or even 25 years, although the wires may very likely be preserved, and some engineers avow they are. The failure of a suspension bridge at Angers, many years ago, showed that it was impossible to keep the hydrate of lime coating there used in immediate contact with the cables, the consequence being that moisture, and other corrosive influences, reached the metal, and the coating of lime was thus rendered practically useless.

In some recent French wire-rope suspension bridges, at the anchorage of the cables, an alloy of 4 parts tin, 5 of lead, and 1 of antimony, has been used for filling between the wires in the conical hole instead of lead. It is found that very fusible alloys are too soft to form a solid anchorage; and alloys which fuse only at a high temperature weaken the wires, and to obtain a long continuous wire it is necessary to unite several separate lengths. Instead of binding the wires together with fine wire, or screwing them into little hollow cylinders, which latter it is found expose them to oxidation, a method has been used of filing the extremities of the wires obliquely, and of soldering them together; the

L

objection to it being that though the joint is strong, the
wire is somewhat weakened by it. In either wire or link
suspension bridges, the anchorage and saddles will be found
to be the places where corrosion has to be especially guarded
against. It is well to remember the larger the superficial
area of the wire cables, the more surface there is exposed to
oxidation.

On September 15, 1886, as a squadron of Uhlans was
crossing a suspension bridge over the Ostrawitza, at Mährisch-
Ostrau, the bridge gave way, and many were killed or
injured. It was commenced in 1846, and not finished till
June 1851, owing to floods. The fracture occurred in one
of the anchor chains. It was found that the material was
thoroughly altered in character, the iron being oxidised
through, so that it could be crushed in the hand. The
anchor stays in question consisted of 12 links, of which one
was completely corroded; the others were reduced to about
¼th of the original section. The original sectional area
of the chain was 24·4 square inches, but it rusted away to
about 4 square inches. The chamber enclosing the portion
of the anchor chains where the fracture occurred was open
to all the surface drainage of the road. A somewhat peculiar
circumstance connected with this failure was that, in July
1885, an official inspection and report had been made, at the
urgent request of the municipality, which said "the bridge
has been examined in all its parts, and is in good and safe
condition."

The ends of the wires are sometimes placed in a tapering
or conical metal tube, into which a pin can be screwed so as
to prevent the end of the wire moving upwards. To the
merits of the system of turning back the strands and splicing
them so as to form a thick end, firmly hold the ends and so
distribute the strain by screwing up, no reference is here
made ; but it is pointed out that unless water, air, dirt, and
moisture are prevented from percolating down the strands,
the wire inside the conical piece may soon corrode, and the
fastenings become ineffective. The Niagara Suspension

Bridge was opened for railway traffic in 1855, and in 1877 Mr. T. C. Clarke, M. Inst. C.E., examined the cable near one of the shoes, and found some of the outer wires corroded through, but the second layers sound. The cause of the corrosion of the outer wires was considered to be that the elongation and contraction of the strands, under moving loads, had loosened the cement from the outside strands, and therefore allowed moisture and air to penetrate. To preserve the wires from rust, they were originally covered with a thick coating of hydraulic cement. Where the hydraulic cement surrounded the wires, they were found to be in a perfectly sound condition, and no signs of oxidation were visible.

In the specification of the steel cable wire for the East River Suspension Bridge, drafted by Mr. W. A. Roebling, the following clause referred to galvanising the cables.

" 8. The cables of the East River Bridge are suspended directly over a salt-water stream, and are, in addition, exposed to the salt air of the neighbouring seashore. Experience has shown that the ordinary means of protection, such as paint, oil or varnish, which would be ample in the interior, are totally inadequate to prevent rusting in localities so near the coast. The only certain safeguard is a coating of zinc, which acts by its absolute air-tightness, as well as by its galvanic action, and is not easily abraded.

" The galvanising must be done throughout in a thorough and perfect manner ; each ring will be inspected in this regard by the inspector when he tests the wire. All rings will be rejected which show spots imperfectly covered, or are full of rough lumps, showing a defective stripping. The galvanising must be of uniform thickness, and must not scale off, or show any cracks when the wire is bent.

" The attention of the manufacturer is particularly called to the point that he must galvanise at such a temperature, and in such a manner, as not to destroy the temper of the wire.

* See ' Scientific American,' 1881.

The manufacturer must run the whole risk in this respect, because the wire is inspected and tested after it has been galvanised. Samples which have been received and tested show that it is not difficult to reconcile these two operations, and that when proper caution is exercised, and the parties possess sufficient experience, the wire can be properly galvanised without impairing the temper."

Mr. F. Collingwood describes * the results of an examination, in August 1883, of the Alleghany Suspension Bridge at Pittsburgh, U.S.A., erected and in constant use since 1861. Each inner cable contains 2100 wires laid up in seven strands, and measures 7½ inches in diameter. Each outer 700 wires laid up in two strands, and is 4½ inches in diameter. The serving or wrapping wire on the cables measures 0·098 inch in diameter, and is included, of course, twice in the diameter given for the cables. The interior spaces were filled in solid with hot coal-tar, which had been boiled and treated with quicklime to neutralise any acid it might contain. It was found, in some of the subterranean portions of the cable, which had in addition been covered with canvas, that the tar had partly disappeared, and that the cavity was nearly full of a dirty-greyish liquid. There was also extensive rusting of the wire, so that the seizing wires, 0·06 inch in diameter, were in many places rusted through, and the cable wires deeply pitted. A second cable-end was opened with similar results. A general survey of the bridge revealed fine cracks in the paint on the cables, which admitted moisture, &c. Recommendation was then made that all the cables and other ironwork should be scraped and repainted. The serious damage to the wires was found to extend about 3 feet from the anchorage outward. Beyond this there was a little dry rust, but no pitting, and still further from the anchorage the paint on the interior wires was gummy.

The rust appeared to be of two kinds : a red oxide where the wire seemed to have been attacked as if by acid; the

* 'Minutes of Proceedings,' Inst. C.E., vol. lxxvi.

second form of rust was a hard blackish substance, containing much sulphur, which, when scaled off, left a deep pit. The result of a chemical analysis showed the rust to be a com-bination of the hydrated peroxide of iron and sulphate of iron. The liquid found among the strands consisted of a weak solution of carbonate and sulphate of ammonia, coloured by tarry matter, and was almost identical with tar water from gasworks.

The original cables were first coated with boiled linseed oil, and afterwards with coal-tar. It was remarked that the tar had evidently not been heated long and high enough to drive off all the water, and the salts of ammonia, contained in coal-tar. Professor Wuth, of Pittsburgh, further reported that " the oils contained in the tar first dissolved the coat of linseed oil ; then the sulphuret of ammonia, which is contained in the tar in considerable quantity, acted upon the iron, con-verting it into the sulphuret of iron, which again was converted into the sulphate by the oxygen of the air, which could not have been completely excluded. This alternate action of the sulphuret of ammonia and tar was continued until the sulphuret was entirely exhausted. The oxidation was further carried on by the atmospheric air in the presence of water and carbonate of ammonia." The presence of water was accounted for by air having slowly percolated to and fro as the masonry changed in temperature, and moisture would probably be condensed, and the water slowly collected. The sulphur and ammonia, it was considered, accumulated in this way, as Pittsburgh is a very smoky city. The preserving measures adopted were, first, all the wires were cleansed thoroughly by scraping, as drenching was found to be im-practicable. Wedges were used to force the strands apart. It was found the damage was almost entirely confined to the outer two layers of wires in each strand. The seizing wires had held the strands so close together as to exclude the destructive agent from further penetration. After the wires had been repaired, by jarring with mallets all loose rust and dirt were removed, and the wires were thoroughly saturated with raw linseed oil. Two days afterwards, a coating of boiled linseed oil was applied, and then the strands were tightened up by the seizing wires. The cable was then saturated with white lead and oil. The cable-end for some two feet back from the clamp, the length being surrounded by canvas, was bathed in melted paraffin until it could hold no more. Changes of temperature had, on examination some months after, had no effect upon the paraffin as a protection.

It would appear that, although wire suspension bridges are excellent for sustaining strain, unless corrosion is thoroughly guarded against, the solid link or rod-suspension bridge is to be preferred. It is open to question whether a linseed-oil coating is a durable and effective coating.

Anchor blocks of Portland-cement concrete are to be preferred to those of masonry or brickwork, as the former can be made a monolithic mass, but the latter have joints of uncertain and unequal strength, and impervious Portland-cement concrete is a proved preservative against corrosion ; masonry and brickwork not only do not always so act, but may be the cause of the introduction of active corrosive influences, either along the joints, or through the stone or brick.

The anchor cables formed of 127 wires, 0·19 inch in diameter, on the right bank of St. Christophe Suspension Bridge,* it was found had been so reduced from the oxidation of the lower portions of the wires, by the oozing of fresh water into the gallery, and the intermittent action of sea water, that they had to be renewed in 1884. Ten per cent. extra thickness of wire was allowed to provide against corrosion at exceptionally exposed places. Experiments on the portions cut off from the old cable showed that the wires erected in 1850 were uninjured, and had lost hardly any of their primitive strength.

In the Cincinatti Suspension Bridge, so far as oxidation is

---

* *Vide Annales* des 'Ponts et Chaussées,' 1886.

concerned, all the parts are open to inspection. The wires composing the cables are well protected by varnish, and so closely and compactly compressed that all interstices are filled with linseed oil. This protection is further increased by the outside wrapping, so that it was considered no apprehension need be entertained as regards corrosion. On taking apart the cables of the suspension aqueduct at Pittsburgh, after seventeen years' exposure, and the effect of dripping water leaking from the trunk, and with little or no care bestowed upon them, the wires inside were found just as free from rust as on the day they were put in. The cables are wrapped from end to end with No. 10 wire, which gives them the appearance of solid cylinders. A well-manufactured cable, it was considered, may be regarded as perfectly free from all danger of rusting; however, the examination of several wire-rope bridges, previously mentioned, hardly confirms so sanguine an assertion.

CHAPTER XI.

NOTES ON THE CORROSION OF METAL EMBEDDED IN CONCRETE,
BRICKWORK, OR MASONRY.

WITH regard to metal embedded in concrete, brickwork, or masonry, it is not advisable to assume, when iron or steel is embedded in brickwork, masonry, or concrete, that moisture and air are excluded from it, and that if it be cleaned and tarred or painted before being covered then it cannot corrode ; for vibration and weather influences cause the mortar in the joints to perish, and cracks occur, and this may even be the case with good Portland-cement concrete. When a fissure appears it is a conduit by which moisture, air, and other corrosive influences can reach the metal. By pointing the joints of brickwork or masonry, as required, and closing any cracks or fissures as they occur in Portland-cement concrete, such deleterious action can be lessened. However, if ironwork is free from any corrosion when placed in position, and properly cleaned before it is coated, and is fixed in air, damp-proof, and water-tight concrete, brickwork, or masonry, it is unlikely to appreciably corrode, except under peculiar circumstances—such as the material it is embedded in being impregnated with, or containing, some corrosive influence acting in conjunction with galvanic deterioration of the metal. Brick and stone are permeable and porous, and, moreover, retain some moisture, quite apart from consideration of the limiting zone of capillary attraction from the ground. Stone, such as Portland, will absorb water to a considerable amount. For instance, in immersion in distilled water, some experiments made by Mr. Spiller, F.C.S., showed that the "Whit bed" Portland, used for external purposes, absorbed 92 grains in 1½ hour, the stone weighing 1421 grains, or 6½ per cent. "Base bed" Portland, fit for internal decoration only, being so porous, absorbed 126 grains in 1½ hour, the stone weighing 1291 grains, or nearly 10 per cent. Underburnt, dusty, soft bricks, soon decay in damp situations. Bricks which are dense, hard, even in texture, and have a vitrified appearance, will resist decay. The weathering qualities, density and uniformity, should be considered, and that some bricks contain a large percentage of soluble salts. Porous, open-grained stones or bricks must be capable of receiving more moisture than those which are closer grained or denser. Although they are frequently called solid, their weight, as compared with the metals, to some extent indicates that there are interstices between the grains or atoms which may contain moisture or air, &c. If any material consists of many different ingredients, each has its characteristics, and therefore there may be causes of decay, disintegration, and corrosion that would not be present in a less compound material; there may be mechanical action caused by the different rates of expansion and contraction, as well as the chemical action.

Salts are formed in the bricks or stone by decomposition, which implies a degree of decay, and, particularly in brickwork, causes efflorescence, i.e., a white powdery substance to appear, consequent upon the salt losing its water of crystallisation. This is believed to be due to the sulphuric acid from gas-burners and coal-stoves, in the presence of air and moisture, acting upon the silicates of lime and magnesia in the brick clay, sulphates of lime and magnesia being formed in the bricks. The bricks becoming wet, the solution evaporates, and efflorescence occurs. Some aver it is due to a minute vegetable or fungoid growth. However, no matter how such an effect is caused, it is decidedly corrosive in its tendency, and should be guarded against as much as possible. This efflorescent and corrosive influence is greatest between any portion of a structure liable to be alternately dry and wet, or humid. Therefore, above the ground, and within the range of tides, i.e., where alternately exposed to air and to submersion, such influence is greatest. As the cost is so trifling, the ends of all metallic girders or joists that must be embedded, and therefore cannot be inspected, should be completely surrounded by Portland-cement concrete of an impermeable character. For this purpose, as impermeability, and not strength, of the concrete is particularly required, a 3 of fine dry clean sand to 1 of Portland cement, or a 2 to 1 mixture, can be adopted, a poorer concrete not being suitable. The ends of a girder or joist should be coated with approved tar-asphaltum composition, or other acknowledged effectual anti-corrosive paint, before being embedded. Some recent experiments by Professor Bauschinger showed that the adherence between iron and Portland cement is as much as 625 lbs. per square inch. It must, however, vary according to the quality of the Portland cement, and nature of the surface to which it adheres. In some experiments conducted with the view of testing the Monier system of iron wire and rods embedded in concrete, the adhesion between the iron and concrete appeared to be perfect, and it was concluded they could be considered a homogeneous mass. Still, much depends upon the dryness and cleanness of the iron, and the adhesiveness and imperviousness of the Portland-cement concrete being uniform and of very considerable amount.

The system of dowelling joints run in Portland cement is much to be preferred to any method of iron cramps for stonework, as the latter not only have an ugly appearance, but soon corrode, become bent, and have a destructive effect upon the stone. When iron has to be placed in contact with any other substance, its porosity, and the capacity the material possesses to contain water, should be considered, for they will in great measure determine its corrosive influence, disintegration by frost, and the presence of deteriorating agents.

Iron embedded in properly-made and mixed water- and

air-tight Portland-cement concrete, has not yet been shown
to rust, and the preservative effects of such concrete may be
considered to be established, provided the surface of the
metal was clean and dry on the Portland cement coating
being applied, and free from corrosion ; and, as the expansion
of cement and iron by heat are nearly the same, there is no
struggle between the substances to cause cracks, fissures,
scaling or disintegration.   One of the most conclusive proofs
of the value of Portland cement as a preservative against
corrosive influences is, that formerly it was the practice, in
the case of iron vessels, to coat the inside surface of the plates
with a Portland cement fine gravel and sand mixture, and it
was found to last as long as the ship, the metal almost
invariably being completely preserved from rust.   It has
now given place to bituminous and other coatings, simply
because they are lighter.   Other proofs of the value of
concrete in preventing corrosion may here be mentioned, for
a high authority, Mr. T. Curtis Clarke, has stated, in a report
on the Niagara Suspension Bridge, that the painted anchor
links of that bridge in 1873, eighteen years after its erection,
which were embedded in hydraulic cement, had not rusted,
and the paint on them was found to be fresh, and the seven
strands forming the southerly pair of cables, so far as they
were exposed to view by the removal of the cement surrounding them, were in a perfectly sound condition, no signs
of oxidation being visible.   In another place, that of the
cable wires, which, of course, were not embedded in cement,
the paint had cracked along the spiral lines of twist, admit-
ting water which caused rust.   It has been found that
water-tight Portland-cement concrete preserves the buried
part of a telegraph post from corrosion, and also fulfils the
double function of ensuring its stability.   Sheet-iron posts
not embedded in water-tight Portland-cement concrete soon
corrode, and the portion above the ground has to be painted
every two or three years.   Ironwork which, it is believed,
was inserted in the walls of buildings erected 500 or 600
years ago, has been found, when embedded in lime mortar, to

be in a fair state of preservation.   On the other hand,
M. Eiffel has stated that some iron rag-bolts, which had
been placed in fortifications for 200 years, had become enlarged to from 2 to 2½ times their original diameter by
rusting in mortar.   Probably this was due to oxidation of the
metal having commenced, or to its being damp or dirty when
placed in the structure.

157

CHAPTER XII.

THE INFLUENCE OF DESIGN AND WORKMANSHIP WITH REGARD
TO CORROSION.

The influence of design, so far as regards corrosion, is very
considerable, and the serviceable life of a metallic structure
may be greatly prolonged by attention being directed to
prevent the decay of any of its parts without detriment to
the best distribution of the metal to meet the strains. Simplicity of construction is one of the chief objects to be
desired. However, it may not be easy, or possible, to design a
structure so that every part can be inspected and maintained;
still, if any portion is inaccessible, so that it cannot be coated
and protected from time to time, corrosion must take place,
and increase. But much can be done towards the accessibility
of the parts by simplicity in the form of the various members,
and by care being exercised that complicated arrangements
are avoided as far as practicable, and by all depressions in
which water may collect being drained by holes, or filled with
a durable waterproof mastic.

In bridges, although there may be disadvantages in
having as few members or separate parts as possible—because
of deflections, bending strains, vibration, greater importance
of accurate and perfectly strong joints, inequalities of material, and faults and flaws, &c., &c.—possible corrosion is
undoubtedly lessened in the case of thick, massive sections, as
compared with thin and light, which have a larger surface
area exposed to corrosive influences as compared with the
heavier and thicker members.

Wherever bars or pieces cross, or are bolted or riveted

together, water permeates into the interstices, and deflection and vibration of the parts of a structure, caused b; traffic or shocks, accelerates any corroding action ; therefor(it is well to reduce to a minimum the joints and surfaces n contact, the effect of a rolling load being sometimes to rais(, and then to deflect a part of a girder. As a metallic structue consists of a combination of a number of members, none of whose dimensions are usually nearly equal to its length, ınd few to its height, the joints are of great importance, and, ıs they are weakened from unequal bearing, and other defecs, a little extra strength will not cause an excess of metal. Corrosion, when accelerated by galvanic action, acts with tle greatest energy in the interior surfaces of joints and such iıaccessible places. Every member should be designed so tlat it may, relatively to the strain it has to bear, be as strⱦg as any other part, and be free from corrosion ; for surplɥs strength may cause local weakness, and result in portions of ı structure being especially liable to corrosion and deterioraton. Continuous girders, particularly, may be brought t(a serious condition consequent upon looseness of the parts that may not be apparent, and by corrosion ; and there is always more risk in their case from any deterioration of tıe section.

In metallic structures, in addition to the metıl required to meet the strains, the contingencies of manufaⅰture, construction, vibration and corrosion, and any otıer special circumstances, have to be considered. It may bⱦ advisable to design them, even when modifications in theiı theoretically correct form have to be made, so as to allow oⅰinspection and sufficient space for painting and repairs. ⸴f possible, every part should be so designed that it can be examined, repaired, and painted, without danger to life ; anⱦ it may be expedient to provide special facilities for thesⱦ purposes, such as examination-platforms, movable ladders, ⸴c. There is no advantage in adopting other than the usua and most suitable forms of iron, if merely with the vier to lessen corrosion, when it can be appreciably prevented ɬy ordinary care and attention, for they have satisfactorily withstood the test of many years' use. The object in designing ironwork, apart from questions of strain, &c., should be to expose as small and thick a surface as possible to corrosive influences, and allow of easy access to every portion of a structure. Practical experience may often overthrow any purely scientific theory of the best shape or section. The strongest forms are generally the simplest, and the fewer the parts the better, or increased facilities for corrosion will be provided. A section or form difficult to roll or cast is not so strong as one that can be easily made, or without much difficulty. It should be remembered that when sections of iron are wide, as well as deep, they are not easily rolled, nor is a channel-iron of considerable width, or a deep wide H-section. In the case of iron of a plastic and inferior nature, such beams are not nearly so difficult to roll. Z-bar sections, deep angle-bulbs, L and T-bulbs, it is said, can be more easily produced in steel than iron, and lengths of 60 to 70 feet have been made. This saves joints, and lessens corrosion. The cellular form of member is not often now used in girder-work in Great Britain, because it cannot be painted internally or repaired. If a box girder, column, or any cellular section, has to be adopted, which cannot be inspected, it is well to line it with Portland cement, or to fill it with comparatively rich Portland-cement concrete, so as to protect the surface from corrosive influences, such as air, moisture and vapour. The form of a piece of iron or steel has, by careful observation, been found to have an effect in preserving it from corrosion, and it has been shown that pointed or square edges, abrupt bends and angles, particularly in submerged, or occasionally submerged work, should be avoided, and that round and curved surfaces are less subject to corrosion. Whenever various forms and sections are used in a metallic structure there must be much manipulation in order to make them fit. Imperfect fitting and joints must then increase, and corrosion ; the strength of a structure be diminished, and be more liable to deterioration.

The limit of elasticity is the chief point to ascertain in

determining the safe strain on metals, for it should not be
deduced from the breaking-weight alone.   To know the
strain which the material will bear without injury to its
elastic powers is absolutely necessary to enable a correct
design to be made.   If every care is taken that corrosion
cannot occur, or is at once arrested, and any strain brought
upon it is well within the limits of elasticity, the metal
should not deteriorate seriously ; but whether age and being
in one position do cause deterioration has not been authorita-
tively determined ; however, it has been shown that strained
and unstrained iron do not equally corrode consequent upon
galvanic action.   The stretching of a rod, chain, or bolt,
when tensilely strained, is not uniform unless the mass of the
metal is of the same strength.   The greatest extension will
occur at the weakest part.   In compression members it is
doubtful if the metal always gives equal resistance, for that
portion which is situated at the greater distance from the
neutral axis is in a position of increased resistance to strain,
as compared with that which is in its immediate proximity ;
therefore, unequally-sided sections are not generally subject
to the same strain throughout their sectional area.   The ties
of a bridge should never be cut or nicked, but be left entire.
If they have to be joined, corrosion at the joints will be a
very serious matter.   The best and most careful workmanship
cannot make a constructed section of iron equal in strength
to the same section of the same quality of metal as it issued
from the rolls.

In some experiments made by M. Dupuy, with apparatus
for measuring strains, it was found that, as the braces or
lattices bear only a portion of the shearing strain, and the
remainder has to be borne by the flanges, it is important
that the flanges should not be greatly reduced at the abutments, where the shearing strain is greatest.   The strains
also produced on the braces of the same panel are not equally
distributed between the braces in tension and compression.
and the flanges have a tendency to overturn, hence any
corrosion of the metal in the flanges at or near the supports

should be especially prevented, corrosive influences being frequently active about the ends and portions of girder-flanges near piers or abutments.

Some flaws and defects of workmanship are likely to occur in built-up metallic structures, and when to these corrosion is added, a serious reduction of strength inevitably follows, which, sooner or later, will affect the stability of the whole structure. Apart from the question whether the same sectional area of metal, if in *one* piece of iron, will bear without fatigue more strain, or the same strain, with less deteriorating effect, than if it were in two or more pieces, it is a protection against corrosion to have all parts of considerable sectional area for the reasons previously stated in this chapter.

No metallic structure, when the sun shines on one side of it and not on the other, and the undersides are always in the shade or exposed to cool winds, is at rest, for there will be strain from this cause, and it will tend to loosen the parts, because of the unequal expansion and contraction, and one side or part may be more susceptible to corrosive influences than another. Thus, the top and bottom members of a girder may be unequally exposed to the sun's rays, and also one side receive almost all the sunshine, and so any tie-rods and bracings may work loose. Alterations of weather have not such an effect, as they are generally felt throughout a structure. For these reasons, although a plate girder of considerable span affords a cheap screen or parapet, it is doubtful if it is as durable, even when carefully maintained, as an open-web girder, for there will be a tendency for moisture to be attracted from the colder towards the hotter side, and the paint—which in order to increase radiation, and diminish the effect caused by alterations of temperature, should be white, or nearly so—is more severely tried on one side than the other, according as moisture or dryness may deleteriously affect it; therefore, corrosion is likely to be unequal. If one side of a bridge or structure is shaded, the temperature of its surface may be less than that of the air, and if the latter is in a saturated condition, a deposition of moisture will take place, and any corrosion be accelerated. Those portions of a structure liable to be shaded, or in a dark and partly enclosed position, are especially likely to be corroded. It is advisable, in designing, to avoid troughs and channels in which one or more surfaces are shaded. Although the less surface exposed to the air per unit of sectional area the better, that which in the nature of things, when properly painted, must be exposed, it is well to expose to the air as much as possible to counteract dampness; and also to avoid all forms and sections in which water can collect, or dust, earth, or vegetation, come in contact with the coated iron.

Open-web girders, such as the lattice, Warren, the various trusses, &c., have one advantage so far as regards preservation from corrosion, namely, that the sun and air reach them more uniformly, it being well to expose the metal to the air, unless it can be hermetically kept from it. In designing lattice, truss, or any form of girder bridge, if considerations of strength, rigidity, riveting, and bearing as a strut do not prevent it, the thicker a bar, and the less the surface exposed to the air and weather, the better; and should light bracing be necessary, if it can be done so as to effectually brace the parts, without too great concentration of the strain on the principal members, it is advisable, considered from the point of view of corrosion, to *reduce* the *number* of bracing pieces to a *minimum*, so as to have as few joints as possible, and thicker bars; and also in lattice girders, or any in which bars cross, to have as few intersecting members as possible, because moisture and dust, &c., penetrate and accumulate at the points of crossing, and the vibratory action caused by a rolling load helps to loosen the joints and aids corrosion by mutual pressure of the parts.

All plates, angle, T, or other pieces of iron, except where especially shown or required by the drawings and specification for the purposes of camber or otherwise, should, by preference, be perfectly straight or regularly curved, so that no depressions are formed for the accumulation of dirt, dust, moisture, damp air, smoke, earth, vegetation, or gases, &c., which will quickly cause corrosion to actively proceed ; and every angle, bend, and corner, should be so arranged that it can be inspected, cleaned and painted, and care should be taken in the design to prevent water finding its way down any joint, or between plates, or any contiguous members of a structure.

As pressing iron into any reasonable shape is now so often done while it is hot, and without the injurious effect of hammering, there may be cogent reasons for using bends, curved, and other forms.   However, there is no occasion in adopting them to do otherwise than avoid depressions or receptacles in which corrosive influences can accumulate. Apart from consideration of strain, care should be taken that all flat bars are perfectly straight, and that they are not slack, or moisture and dirt are certain to collect, and corrosion be accelerated.   Packing-pieces should be avoided as much as possible, as they are not always shaped or planed so that they fit tightly, and they should be of the same metal in every respect as that with which they are to be in contact, and not always of cast iron.

One of the most important provisions against corrosion is that all plates of a built-up metallic structure should fit so tightly, and be in such contact, as to exclude all air, smoke, vapour, moisture, water and dirt ; therefore, a closely-riveted structure is likely to be less affected by corrosion than one with more widely-pitched rivets, always provided that the edges of the plates are flat, and not turned up or distorted. To prevent air, smoke, vapour and moisture, from penetrating between the plates of a girder or structure, it has been suggested that preservative material might be forced be- tween the plates by a powerful syringe.   It is certain, if any preservative could be made to permanently fill such vacuities, it would be more efficacious than any covering of the edges of the plates.   Close riveting, if not weakening the effective area of the plates, tends to exclude corrosive influences, pro-

M 2

vided it does not damage the edges of the plates, or cause them to turn up or bend.

Thick plates are usually stronger per unit of area than thin plates. If it be necessary to use a thin plate in building up the flange of a girder, it should be placed at the bottom, or it will be liable to buckle. Although it may be advisable to have the thickest plate in the top of any flange, as it is considered the strain is greatest there, plates of equal thickness are, on the whole, generally preferred. Experience has proved that which was to be expected from the less exposure : namely, that plates do not corrode so quickly if they are united in a considerable mass, or if a thick one forms the bottom or top flange of a girder, provided the joints and riveting are water and air-tight. Sir B. Baker's reliable experiments have shown that it is best to make the thickness of the plates of flanges of uniform size, and that narrowness of the bottom flange may cause a girder to fail before the real strength of the metal is brought into action. If desired, the width of the top flange, when the load is on the bottom flange, can be made narrower, and the thickness increased, but the bottom flange should have a reasonable width ; and rigidity of the flanges, and web, which is generally considered as being subject to diagonal compression due to the shearing strain, should be assured. In so designing a flange, attention can be given to the prevention of corrosion by having no thin plates. The strength of a wrought-iron plate is generally not uniform. In the middle, where the puddle-bars forming the pile are most drawn out, it will probably be greatest, and decrease gradually towards the edges. The designed width of a plate-girder bridge varies in different countries. In this, wide flanges are adopted to ensure stability, and to obviate the necessity of overhead bracing. In America narrow flanges, compared with ours, are used—say, from 0·75 to 0·50 of the width here adopted. Of course, with a narrow flange, thicker plates are required to make up the necessary sectional area of the metal, but in any than short-span plate girders for light railways, the area of metal required causes a $\frac{3}{8}$-inch or more plate to be used; consequently, any apparent advantage the narrow-flange system possesses, on the ground of thicker plates and less possible corrosion, because of the less surface exposed to corrosive influences, does not then accrue. It is an advantage if all the plates are of the same pattern, and to use one thick plate in preference to two, and none less than $\frac{3}{8}$ inch in thickness, and to build up any section with the least number of plates that due consideration of strain and manufacture may require. Although $\frac{3}{16}$-inch plates are used in plate-built structures, it is better to have them not less than $\frac{1}{4}$ inch where only thin plates are required to give the necessary strength. It may be generally accepted that in built-up plate flanges it is almost impossible to make them fit so closely as to prevent the ingress of air, smoke and vapour; and even if this can be done, when such a girder is erected, vibration, strain, deflection and set, &c., will prevent them being an absolutely air and water-tight mass. The lower flanges and parts of girders are particularly liable to become quickly corroded, for water, dust, and vege-table growth, ballast, earth, and the drip of locomotives, are all, more or less, there concentrated. Around the angle-irons, junction of bars, cross-girder bearings, all joints and connections, rivet-heads and holes, and any part of a bridge in which water or dust can accumulate, the point of attach-ment of ⌶-iron and floor-plates to girders, where any timber is in contact with the metal, as in the wooden platforms of bridges, corrosion is to be expected, because of the presence of tannic and gallic acids; and the wood will also suffer if placed on *bare* iron, especially in tropical climates : in fact, all places that can be receptacles for, or are exposed to, the influences mentioned are particularly liable to corrosion.

If rust has loosened the attachment of rivets, angle, ⊤, channel, or other pieces of iron, to a flange-plate, there is no uniformity of strain, and each part in turn may be tried severely instead of all acting together. The late Mr. Cowper, M. Council, Inst. C.E., said with respect to built-up girder-work, "He would suggest a mode of protection should be adopted, by forcing varnish or boiled oil, or whatever might be the best material, between the plates by a powerful syringe ; the oil might not go all the way in, but it would go as far as the air would, and two or three injections might entirely stop the entrance of air. In that way he believed many bridges of laminated plates, which were now in a state of deterioration, might be preserved from further deterioration between the plates. Thirty years was too short a life for a bridge, and some improved method of protection ought to be adopted."

It may be well to remember, in designing, that rough and uneven, and polished surfaces, of the same metal, will not equally corrode, because it is not easy to make water adhere to a polished surface—as, for instance, to a needle or a razor—until it becomes dirty, and it then finds a place to rest, as no indentations or cavities are present to contain any rain-water or moisture, and the area exposed to corrosive influences is less than that of a rough surface ; for it is an axiom, which cannot be too much borne in mind in designing metallic structures, that the greater the extent of surface exposed to atmospheric influences, the more powerful will be the action of air and water to corrode it.

In designing, what additional thickness should be allowed for corrosion ? It principally depends upon the active power of the agents of corrosion to which any structure will be subject. These must necessarily vary according to the location, material employed—whether steel, wrought or cast iron—and all the other influences that promote corrosion. The amount of the corrosion of cast iron in a century, when placed in clear sea-water, has been variously estimated—by deductive reasoning from experience, observation, and the results of tests during a few years—at from $\frac{1}{10}$th to $\frac{4}{10}$ths of an inch, according to the quantity of graphite or carburet of iron in it; white iron being considered to corrode the slowest. The corrosion of wrought iron, under the same conditions, was estimated at from $\frac{1}{10}$th to $\frac{6}{10}$ths of an inch. However,

such general estimates are unreliable, for the varying
quality of the iron or steel, local conditions, and the intensity
or comparative absence of corrosive influences, render any
such deductions almost worthless, for in some places iron has
been very seriously corroded in a very short time, reference
to which has been made in Chapter VI.; on the other hand,
in some structures, it hardly appears to have commenced.
To follow any attempted formula, derived by generalising
from a few examples, and declaring a fixed rate of corrosion,
is to be subject to a false guide.   Each case and structure
must be separately considered, and every circumstance be
duly noted; and that the action of corrosion generally
becomes intensified by age.   Perhaps the time will come
when it will be the custom to add to the sectional area of
the metal required to meet the calculated strain a certain
percentage to allow for corrosion, and this has been done in a
few cases; for, wrought iron and steel, when very thin, are
much more liable to waste away, from corrosion and the
effects of vibration, than fail from the strains brought upon
them.   Mr. Shelford, M. Council, Inst. C.E., in the designs of
the iron bridges of the Hull and Barnsley Railway, carefully
considered the question of durability as against first cost
cheapness, and made the sections thicker according to the
exposed situation of the bridge, so as to reduce the cost of
maintenance.   One-sixteenth of an inch in thickness is, in
some American specifications, allowed when *one* face only of
the metal is accessible.

In all metallic structures to be erected in the tropics,
ample allowance for corrosion and deterioration should be
made in the thicknesses, and all plates and parts should be
of the same pattern, so that they can be easily and accurately
fitted.   It is desirable that any metallic bridge should be put
together at the makers, if possible, as then there is less trouble,
and no risk on erection at the site, and parts can be made
to fit better, and the possibilities of corrosion are reduced.
In the case of underhung cross girders, which rely upon bolts
and rivets to hold them to the main girders, and to keep the

floor of a bridge in position, special provision should be made to protect the connections from corrosion, and to so arrange them that they can be easily inspected and renewed.

As the upper portions of a bridge are the most exposed to carbonic, sulphuric, and sulphurous acids, generated by the fuel, steam, and heated gases of locomotives—chlorine and ammonia also probably being present either from the air or water of locomotives—it would seem that it would be advisable to have as little ironwork as possible *above* the rail level, and therefore to have the load on the top of girders. However, special circumstances, such as headway, must govern each design, and the load on the top of a girder system is not much used in this country, as parapets or screens are then required in many situations; but, considered merely from the point of view of the prevention of corrosion, the system would appear to be preferable, provided the bridge floor is properly designed and maintained, so as to prevent percolation of water to the main and cross girders.

The Russian Ministry of Roads, a few years ago, published some rules respecting the employment of steel in bridges, and one is that iron and steel may be used in the same structure, but in each member of a group of similar parts the same material must be adopted. Thus, the top and bottom booms of a girder, the diagonals and verticals, cross and longitudinal roadway bearers, form such a group. This is a provision against corrosion by galvanic action, but it is always preferable to employ only one metal, and every care should be taken to ensure that it is homogeneous and of the same composition. Sir G. B. Bruce, in his address, in 1877, as President of the Institution of Civil Engineers, expressed the opinion that flat iron girders for ordinary road, river, and accommodation bridges, not for bridges of large span, where a brick or stone arch is admissible, are greatly inferior both as regards appearance and durability, and, he asked, " Where would the works of past centuries have been now had they been of iron ? "

As a metallic viaduct becomes older, the cost of maintenance will increase, as compared with one constructed of stone or brick, and the charges for maintenance of an iron or steel bridge may make it advisable to construct it of stone, brick, or Portland-cement concrete. The extra strength of steel enables a lighter section to be used ; there is, therefore, a greater area per unit of section exposed to atmospheric and other corrosive influences, hence an even increased necessity for protection against corrosion.

In the new Tay Viaduct, to provide against the effects of possible corrosion, the outer wrought-iron casings of the cylinders of the piers only extend to low-water level, which is about fifteen feet below high-water mark. Inside them is a ring of brickwork surrounding the concrete hearting, of sufficient thickness to encase it if the wrought-iron skin, which was adopted for convenience of construction, should perish.

Great care should be taken that any holding-down bolts are kept free from corrosion, as the stability of a pier or structure, particularly when it is severely strained, may almost depend upon their resistance. In order to enable their lower ends to be inspected, a small subway beneath their anchorage can be constructed, of sufficient size to enable a man to get along it and do repairs, the dimensions increasing according to the length of the tunnel; 2 feet 6 inches to 3 feet may be sufficient width ; the height being not less, for a very short length, than 2 feet 6 inches, and increasing, as may be thought desirable, according to the length of the subway. If it can be made reasonably larger, so much the better.

In the Transactions of the American Society of Civil Engineers, vol. xvii., it is stated by Mr. Buck, that when the old rollers of the Niagara Suspension Bridge were inspected, during the reconstruction of the towers, the old saddles, together with the main cables lying over them, were raised from the old rollers by means of hydraulic jacks. They were found to be embedded in cement and iron rust, the spaces between their ends and the ribs of the bed-plate being filled with this cement, which had become so firm that a chisel-bar was required to extricate the rollers. Thus each roller was lying in a trough, and any water finding its way into it could only escape by evaporation. The old rollers appear to have been placed side by side without a roller-frame, with a play of a ¼ inch between their ends and the fixed ribs of the bed-plates. The new rollers were formed with trunnions, running in holes in the side-bars of a roller-frame, whose depth is greater than the diameter of the new rollers, and overlaps the steel-bearing plates above and below the rollers, the edges of the frame being planed to come into neat contact with the edges of the steel plates and the shoulders of the rollers; the latter are therefore protected from water, and as far as possible from dust and corrosive influences. This shows that, in consequence of corrosion, the rollers upon which the ends of a girder may rest require to be carefully protected from rust or they may cease to act as rollers, and, in addition, cause corrosive agents to accumulate, and be active centres of decay.

The bottoms of petroleum-storage tanks, when placed on earthy soil, especially that containing much moisture, soon decay, as they cannot be inspected and repaired; and all tanks should be raised sufficiently above the ground so that the bottoms can be inspected, repaired, and painted when necessary. In the case of plates riveted together to hold oil, the rivets must have a *less* pitch than that necessary to hold water. This is an advantage, as the plates are likely to be held together tighter, and corrosive influences excluded.

The system of Messrs. Moreland of encasing the columns, girders, and cross-bearers of a floor in Portland-cement concrete, apart from the claim that it is fireproof, is to be commended—provided the inside of the columns are filled with Portland-cement concrete after being made perfectly clean and empty—as being an excellent method of preserving the ironwork, the construction being a continuous mass of concrete. It may, perhaps, be found even more effectual in preventing corrosion than in resisting a fierce fire.

In the case of iron or steel dock-gates and caissons, vibration caused by blows of vessels produces leakage when they are constructed on the box principle, but if the water is at the same level on both sides, such shocks are very much reduced in intensity and effect.   In order to lessen corrosion, all parts should be open for inspection, repair, and painting, as they are generally exposed to severe corrosive influences; and therefore, to lessen corrosion, single skin-gates are to be preferred.   It is generally difficult to keep dock-caissons and gates water-tight, as they are usually subject to considerable shaking and jarring, and therefore to somewhat rough treatment.   In order to simplify the connection of the skin-plates, if of varying thickness when designed to resist the strains, it is desirable to consider whether they had better not be made of one thickness throughout.   This can be done by arranging the position of the ribs so as to obtain the required strength.   In making water-tight joints between iron and timber, it is well to remember that a small face-area of timber is easier to make water-tight than a large one.   After having made the timber-end to fit the ironwork, it can be coated with Stockholm tar, or other known preservative, lines of tarred spun-yarn can be placed on the timber-face, and it can be fixed tightly to the metal by screwing up the bolts.   Portland-cement grout can also be used, or tarred felt.   The heads of the connecting bolts can be sunk below the surface of the timber, see Figs. 1 and 2, p. 172, and be covered with neat Portland cement, and the nuts, if possible, should be placed inside.   The holes are sometimes also plugged with greenheart, run in with marine glue.   If a soft substance is used for washers, such as indiarubber, on tightening-up they may turn up or be torn.   The ends of the bolts sometimes have cast-iron washers, with indiarubber packings under the nuts, so as to cause them to be water-tight.   If iron washers are slightly cup-shaped, so as to receive a rubber washer and hold them in place, they are much less likely to be torn, bent, or turned up at the edges.   Hard rubber generally resists chemical action much better than the soft

material.   Indiarubber containing mineral matter becomes
hard and brittle if stored for several years.   Joints are
sometimes made by placing slips of wood, 1½ inch or so
in thickness, between the metals, and by caulking.   As
the wood will be more or less always in a damp state, it
cannot be said to be an anti-corrosive joint.   To obtain
thorough caulking in timber, about 3½ inches of planking is
desirable.   A water-tight junction between two built-up parts
of an iron-plate caisson has been made by a strip of indiarubber ⅜ inch in diameter, laid in a groove at the top of one
portion.   The iron lock-gates of the canal at Havre are
galvanised, and the rivet-heads painted with zinc-white.
The steel lock-gates of Limerick floating-dock were repaired

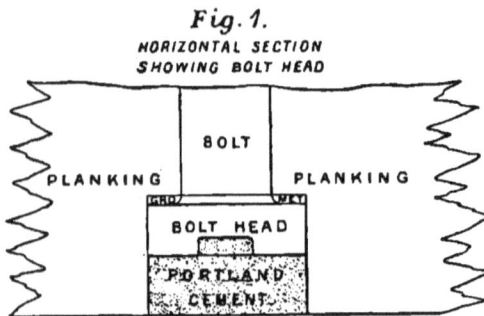

Fig. 1.
HORIZONTAL SECTION
SHOWING BOLT HEAD

BOLT

PLANKING   PLANKING

BOLT HEAD

PORTLAND
CEMENT

Fig. 2.
PLAN OF
BOLT HEAD

by removing all scale and rust from the inside plates, pure
Portland-cement wash being run into all the seams and joints
of the cells, the utmost care being taken that it penetrated
everywhere.   By attention being directed to the preceding
points, and by care in construction, painting, and maintenance,
corrosion in dock-gates and caissons can be much lessened,
and perhaps prevented.

    With regard to workmanship, it has an important effect on
durability and the magnitude of the corrosion.   Care should be
taken that all plates and bars, angle, T, channel, and other
members, are in close contact over the entire meeting-surface,
and that there are no gaping joints; therefore, the ends of
plates, and all exposed edges and abutting surfaces should be

perfectly flat, and be planed, at least, in structures of any importance, and in all, if possible. Planing the edges of plates, so that they are in even and perfect contact, is far preferable to mere caulking, for the caulking-tool tends to separate the plates. Any plates, angle-irons, &c., forming or bearing on the flanges, should be carefully butted, squared, and close-jointed, so that they are in complete contact, especially if they act as struts, and do not depend upon the rivets to take the thrust; and all web-stiffeners should have tight bearings against the flanges and web. In lieu of a better joint, where two uneven cast-iron surfaces, having no permanent load to sustain—as the cylinders of bridge piers—have to be joined, in order to obtain a comparatively even bearing, and to prevent corrosion or water penetrating, and so introducing corrosive influences, it may suffice if they have tarred hemp wrapped on hoop iron around the surfaces.

Rivet-holes should be truly concentric in girder-work. They are usually made about $\frac{1}{16}$ inch larger than the rivet for rivets of the most usual sizes—viz., say between $\frac{5}{8}$ and 1 inch in diameter—so as to admit the rivet when hot, and therefore somewhat expanded, to allow for the scale on the rivet, and want of exact correspondence in the holes, &c. All rivet-holes should be drilled, as punching the plates weakens them by straining the fibre of the metal, and they should be made so that the pieces joined come into close contact. The rivets must completely fill the holes, and have full heads, holding the plates evenly, firmly, and tightly together, and leaving no vacuities or interstices which may cause weeping of water. Unless riveting is very carefully done the rivet-holes are likely to be receptacles from which may spring active corrosive influences. The rivets sometimes have a less pitch in the top flange than in the bottom so as to prevent buckling. They also cause the plates to be so close that moisture, air, smoke, dirt, decaying matter, and other corrosive influences, cannot get between the plates. In some cases it may be advisable to make the pitch of the rivets *less*

than usual ; for instance, a $\frac{3}{4}$-inch rivet, with a pitch as little
as 2 inches to $2\frac{3}{4}$ inches, in order to make the plates or pieces
joined more air and water-tight. Corrosion in girders, as in
boilers, is generally active along the joints, seams, and lines
of rivets, unless they are well protected and the riveting
properly effected. The difference between hand and machine-
riveting, which causes the latter to be so much more reliable,
is that the effect of the blows in hand-riveting is chiefly
confined to the particles nearest the hammer, while machine-
riveting has the great advantage that it communicates the
force to the interior metal of the rivet—which should be
wholly heated and be hot when riveted—and so it completely
fills the hole ; but care should be taken to avoid any bulging
or splitting of the metal, or corrosion will be accelerated.
Especially when the rivet-holes in the plates are badly made,
it has been shown by experiment machine-riveting will cause
a rivet to completely fill a hole, whereas when it is done by
hand, it only partly does so, and vacuities occur, and the
heads may be misshapen and misplaced. Moisture, in time,
will reach such voids, and then corrosion of the rivet and
plates will quickly proceed, and the whole structure will
lose stiffness and rapidly deteriorate, consequent upon corrosion, looseness of the parts, and the mechanical wedge-
action of the badly-fitting rivets. Machine-riveting, from its
uniform pressure and certainty of action, is therefore to be
preferred to hand-riveting. Experiments have shown that
the outside fibres of rivets appear to be more strained than
those near the centre.

It is doubtful whether a number of plates are generally
made sufficiently tight at the edges to permanently exclude
air and vapour. If they are not, they become conduits to
convey moisture, air, smoke, and other corrosive influences,
to the mass. In built-up girders, every effort should be made
to keep the metal from air and moisture by accurately-fitting
work, and therefore it is absolutely necessary the riveting
should also be excellent.

CHAPTER XIII.

CORROSION IN PILES AND COLUMNS.

WITH regard to the corrosion of metallic columns, pillars, and piles, which are so largely used, their preservation from corrosion and decay is of much importance. As their thickness, when they are cast iron, is usually from $\frac{1}{2}$ to $1\frac{1}{2}$ inch, and generally does not exceed 1 inch, any decrease in the strength of the metal is not to be lightly regarded, for, when a column or pile is erected, the interior cannot be seen, and in any case comparatively little can be done to preserve it internally, or arrest any oxidation of the inside surface. Their outer surfaces can usually be inspected and protected, but attention is especially required to be directed to the inner surface, so that, although it is hidden when the column is placed in position, it is known that at least every reasonable precaution has been taken to prevent corrosion and decomposition.

For cast-iron submerged piles, the cylindrical form has, among other advantages, that of having no angles or abrupt bends; accord in the order and direction of crystallisation; greater homogeneousness of the metal, and therefore more immunity from imperfections; and the important one that whatever the direction of a blow the same amount of surface is exposed, therefore any corrosive or abrading effect is likely to be uniform. There is, however, a disadvantage in ordinary hollow columns, because, although they may be periodically painted and maintained exteriorly, their interior surfaces cannot be inspected, and therefore their freedom from corrosion must be to a certain extent a matter of conjecture or deduction from the behaviour of similarly exposed columns under the same general conditions. This being the case— although the circular is undoubtedly the strongest form for the amount of metal—in order that all the surfaces of the column may be inspected, and also, in special cases, the cruciform section may have to be preferred, the latter frequently being the handier for attaching or conforming to the internal fittings of a building. By preference there should be no inside obstruction, and never more than is absolutely necessary in columns, pillars, or piles, unless they are filled with impermeable Portland-cement concrete or protective filling, in order that saw-dust, wood-chips, dirt, débris, and corrosive agents may not prevail.

It is advisable to avoid columns that consist of parts joined together by bolts. If compound piles have to be adopted, it is better to use those which are firmly riveted together with rivets that completely fill the drilled holes, and whose flanges and rivets can be inspected on the outside; always endeavouring to have as few pieces as possible, and to avoid all complicated sections, for they are not only difficult to preserve from corrosion, but their strength depends upon their perfect fitting and joint action, which may or may not be the case, and cannot be declared as certain, the probabilities being that, except where great care is exercised, some parts are not much strained till others have been unduly so. As the age of such a structure increases, the parts are sure to set unequally, and become not so tight as when they were erected, then corrosion and strain will quickly complete the deterioration, and will demand serious attention. By requiring the riveting to be most carefully executed, so that the plates are tightly gripped, and by not having too wide a pitch of the rivets, corrosion of the plates forming a wrought-iron column will be reduced, and the strength increased. Many columns of closed cross-section possess the disadvantage that all their surfaces after erection cannot easily be examined or painted. American engineers especially, in many cases, have therefore adopted angle, channel, and trough-shaped sections, braced with light lattice bars, for the struts of bridges.

Inequalities in the thickness of the metal in cast-iron columns not unfrequently exist, although they can, by care, be cast of uniform thickness; however, the thinner part will usually be the harder. Mr. Hodgkinson found that the *external* portion of a casting, whether hollow or solid, was always harder than that near the centre, the hardness increasing with the thinness, and the difference of strength was greatest in the small castings. Corrosion is, therefore, likely to be unequal, the internal surface being more easily affected by corrosive influences. Removing the skin of a casting reduces its compressive strength. An equal H-shaped pillar or stanchion, either cast or formed of two channel irons, or one of cruciform section, can be painted easier than a circular one, but the strength is considerably less. An angle or T-iron light strut is not only stronger, but more easily painted than a channel or cruciform strut, and has a greater thickness, area for area, and is, therefore, not so quickly affected by corrosion.

Columns are now cast with greater precision than formerly, and uniform thickness is to be attained if the best methods of casting are adopted; still, it is prudent to consider that a deviation of 25 per cent. in the thickness of thin columns is not improbable, and that unequal bearing and loading greatly reduce the strength, therefore all joints should be flat and smooth, exclude moisture and air, and be firmly fixed, in order that the strain may be in the line of the axis; and in estimating the strength, an allowance should be made for continued corrosion and decomposition of the material, because the interior cannot be painted, or even inspected, after erection. White cast-iron piles are sometimes preferred, as they are believed to be less likely to corrode than other kinds, but tough grey metal of specified good quality is generally used. Sufficient time has hardly elapsed, since the introduction of iron piles, for any decided opinion to be given, although inferences may be logically deduced. The

N

forms of cast-iron columns generally used are the circular, the +, the H, and sometimes the ⊔. Columns of cruciform section are occasionally adopted when there is a heavy strain upon them, because their thickness can be visibly and reliably ascertained; and, as all the sides are exposed, it may be said that such a form is easier to preserve from corrosive effects than a circular pillar, the internal surface of which is hidden, but the even distribution of the load is not so easily attained. As it has been found that box girders closely riveted and caulked like a boiler, and thus made water-tight, and properly painted before erection, are most effectually preserved from corrosion, it would appear that wrought-iron columns are to be preferred to cast-iron for positions in which a heavy strain has to be borne, and with the view to obtain resistance to corrosion. Wrought-iron or steel columns or supports are quickly replacing, in modern railway structures, cast-iron columns, owing to the accidents that have occurred through flaws or local weakness in castings. The effective protection of the plates from corrosion will, however, require attention.

It is sometimes assumed, because a column is placed upon a column with an intervening cover to it, that it is air-tight and vapour proof, or almost so, and cannot corrode. This is a sanguine assumption. The corrosive influence of any water, moisture, sawdust, wood-chips or débris, that during construction may accumulate inside the column, has to be considered. In every case the interior should be thoroughly cleared and cleaned, and all slime, dirt, and débris be removed, but care should be taken that the silicious skin of cast iron is not disturbed. The interior of piles is seldom sufficiently large to be painted by hand, and the bottom of a column has, therefore, been closed, and paint poured in and allowed to remain for some time; it is then opened and the surplus paint permitted to run out. This process requires more paint, but less labour is necessary. A method of coating to be preferred is that of Dr. Angus Smith's black enamel for protecting water-pipes, &c., or a boiling coal-tar and asphaltum preparation, applied hot as soon as possible after casting, the castings being dipped in a bath of it for not less than thirty minutes. The interior surface should always be coated with an anti-corrosive composition or paint. All the cast-iron piles of the Morecambe Bay Viaduct are protected by a preparation of tar and asphalt, the tar being heated in a large tank, and each piece of ironwork boiled in the mixture for half an hour. It has been very effective. A solution of coal-tar, lime, and sand, poured in at the top, was used to fill the piles of the Clevedon Pier. At Woolwich Arsenal, some wrought-iron cylinders, forming a pier, were coated before immersion with a solution of coal-tar, naphtha, and resin, no rusting being allowed to commence, as then nothing would arrest corrosion except scraping and dipping the iron in an acid bath, neutralising by lime-water, and afterwards properly cleaning it as has been described.

Frequently the base of a column is simply bolted to the sole-plate at the ground line, a position in which it is exposed to dust, moisture, air, and, may be, water. It would be preferable to make the joint a little distance below the ground line and bed and encircle it in Portland-cement concrete, after being properly coated with tar-asphaltum paint, not only to lessen corrosion, but to stiffen the joint, the concrete having an asphalt floor upon it, or to have the joint above the ground line. If the joint is properly made, and all corrosive influences excluded from the column, convenience and strain have then only to be considered, and the positions of the base and joints are of less consequence. Columns sometimes rest on sheet lead, or even sheet copper, which may cause serious galvanic action. A fine Portland-cement grout, or tarred felt covered with such grout, and allowed to soak into it as much as possible, would be as effectual in obtaining an even bed, and act as an anti-corrosive covering, instead of as one which is likely to be a corrosive stratum. The base of a column, whether held in position by a plate, bedded upon concrete or stone, or placed on sheet lead laid in a recess in a stone, is especially liable to

N 2

corrosion unless protected. When the base-plate upon which a column rests is held by holding-down or anchor bolts passing through it, some means should always be adopted to preserve them, for should the base become unstable, or corrosion be unequal—and it is seldom uniform, from various causes—the strength will be much reduced. The capital is often required to support the base of a column, and also a beam upon which are fixed the floor joists. Corrosion may then occur, not only from accumulation of dirt, but also from the timber resting upon the metal, the acids in the wood having been proved to act injuriously upon iron. A properly-composed tar-asphalt, or Portland-cement mortar cushion, or a layer of tarred felt, soaked in Portland-cement grout, upon the metal, so that the timber does not touch it, would tend to neutralise this action, if it did not prevent it. Mr. Mortimer Evans, at the Craigmore Pier, near Rothesay, adopted the system of casing the iron piles with pipes of fire-clay secured with Portland cement. The iron is considered to be the strength of the structure, and the Portland cement the preservative anti-corrosive element, for it is thought the pier will be as indestructible as any that can be made. On certain coasts, such as the Brazilian, iron piles frequently become covered with oysters to a considerable thickness, and other shell and marine life. If the covering is even and continuous, and sufficiently thick to exclude air and moisture, and no issue from the marine animals dissolves or corrodes the metal, it is a protection, but if corrosion has commenced before they seal the surface, it will continue, notwithstanding such a covering, and it cannot be a reliable protection.

The screws of cast-iron piles should be of exceptionally good metal, and great care be exercised in casting, so as to prevent flaws, air-holes, and other defects, and every means should be employed to make them homogeneous. When this is done, as they are not so easy to bend, or liable to rust away, as wrought-iron screws, being thicker, they are largely used with success. In a cylinder-bridge pier,* as the hearting bears

* Vide 'Cylinder Bridge Piers, and the Well System of Foundations, published by Messrs. Spon, Strand, London.

the weight, the bolts and joints are not so important as in pile structures, the especial object of the cylinder-rings being to enable the pier to be erected in their interior, and to afford a protection to the hearting until it is set, and then to shield its surface from weather and currents. The joints should, however, be caulked, or so made that water cannot permeate, and the flanges and bolts should be inside the cylinder—the hearting will then cover them.

The joints are generally the weakest parts of any metallic structure, and they are particularly so in piles, pillars, and columns, hence they should be reduced to a minimum, and careful provision against corrosion should be made. . Imperfections of workmanship, inequalities of the material and bearing, may cause leaky joints in piles, then corrosion will be internally active. Metal to metal joints are considered to be preferable, but very perfect fitting is required to make an air and water-tight joint. White and red-lead joints may aid corrosion, and they cannot be regarded as perfect. As a provision against corrosion, apart from the other advantages, caulking the joints with iron cement, in order to make them water-tight, is to be commended; but, if it can be effected, a metal to metal water-tight joint is to be preferred, but this requires planing, fitting, or machine-faced ends, &c. Among the means which can be employed to obtain tight joints may be mentioned, all the bearing-places to be truly faced in a lathe or planed, so that the abutting ends are perpendicular to the axis of a pile, the bolt-holes drilled, and the bolts turned. A close fit between the bolts and holes is of the utmost importance to prevent corrosion. In designing columns or pillars, care should be taken that the joints are not so placed as to be specially liable to receive corrosive influences, as at the ground level, and where joists and planking rest against them ; for washing the planking may cause trickling or permeation of water down them, and into the joints and connections. As Mr. Hodgkinson observed, in making his celebrated experiments, that "pillars with rounded ends generally broke at the middle only; pillars with flat ends usually broke in three places, at the middle, and near each

end;" these places, subject to other conditions, are those
frequently strutted and tied. Any corrosion at them is likely
to produce decided weakness, and, therefore, it is advisable to
have no joints at or near them.

The ordinary exposed projecting flanges, fastenings, and
joints of piles for bridge piers, and landing and promenade
piers, appear in not a few cases to have been designed somewhat regardless of corrosive influences. Experience with
such structures has demonstrated that the joint-bolts quickly
corrode ; and it is well to remember, apart from other causes,
that this accelerated corrosion is assisted by the different
electro-chemical properties of the materials used, the columns

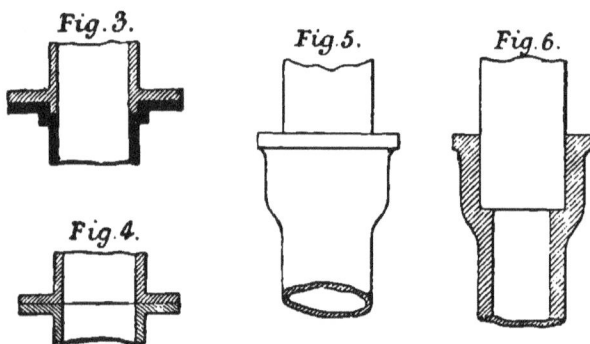

Fig. 3.

Fig. 5.

Fig. 6.

Fig. 4.

frequently being of cast iron and the bolts of wrought iron or
steel. Even if iron cement or caulking is used to fill the
joints or openings, it will generally not remain long perfectly
air-tight and damp-proof. If the only connection between
the different lengths forming a pile, or between a pile and a
screw, are the bolts, when they fail in the latter case, the
column alone must support the whole load, for the screw is
then no longer attached to it. None of these joints or connections can be considered satisfactory, so far as regards
durability and resistance to corrosive influences. If the
diameter of a cylinder or column will permit, joint flanges
and bolts should be internal, and external flanges and bolts
should only be used if absolutely necessary, but some other

more preferable arrangement can generally be adopted. In designing joints of columns, piles, &c., anything which tends to prevent moisture or any corrosive influence penetráting to the interior, is to be preferred; thus, the joint, Fig. 3, is better than the joint, Fig. 4, as in the latter any water or air has a direct course into the column. Figs. 5 and 6 show a simple joint for piles, sometimes adopted to avoid the use of bolts. It is much stronger laterally than that shown in Figs. 3 and 4. The casting is also stronger, as practically there are no angles, and as there are no bolt-holes, the skin of the cast iron is not disturbed. If the junction be made water-tight, it is to be regarded as a good one for the prevention of corrosion.

In order to prevent loose joints and places for the accumulation of corrosive influences, and any mechanical action that may assist to wear away the ends of joints, and so cause openings and depressions, it is well if any bracing to piles or columns can act as struts as well as ties, unless it is *certain* no compressive strain can come upon a tie. The latter should not be cut or jointed. Bracing should have combined action, so as to prevent the strains being concentrated upon one point. The bracing of metallic structures is intimately associated with that of joints, and is necessary to ensure rigidity, and to maintain the required strength, and any deterioration from corrosion cannot but be regarded as serious. Pier bracing exposed to wave action, if of light section, is of very little use for structures in the sea, for it soon becomes bent and loose by the force of the waves, or the constant vibration. For instance, at Westward Ho Pier, 5 inches by ½ an inch T-iron braces were found to be bent by the force of the waves 2 inches out of a straight line in a length of 12 to 13 feet. Thin flat bars should be avoided in bracing piles and columns, as they are much more liable to contraction and expansion, owing to changes of temperature. A much more solid and rigid section should be adopted, as not only being better able to resist strain, but also corrosion, as there will be less exposed surface. Double-headed rails,

56 lbs. per yard, are used at the St. Leonards new promenade pier as struts, and there are no cast-iron lugs on the piles, the bracing being attached to the columns by straps passing round them between two collars, cast on to prevent them slipping up or down. The tie-rods, $1\frac{3}{4}$ inch in diameter, being bolted to the rails, are secured with patent lock nuts. This arrangement allows of inspection, and any corrosion can be seen and remedied. The vibration caused by passing trains on a high, thin column, is considerable, and will have a tendency to make the joints loose. Strong bracing is required to lessen this, and the consequent corrosion. The height of the piers in situations where there is ample headway, can be lessened by causing the load to be on the top of the girder instead of on cross-girders resting on the bottom flange, the height of the piers being lessened by about the depth of the girders.

Bolts passing through any external flanges are especially liable to be corroded, and as the corrosion has been found to be very considerable—even as much as two-thirds of their strength has been destroyed in a short time—it is advisable to do without them, if possible, by the adoption of some other arrangement, such as a slot and stud joint, in which any pressure on the pile is directly transmitted to the thickened end of the flange, or after the manner of the boltless joint just previously described. Sleeve-bracing can also be applied instead of bracing bars joined by bolts or other connections to external flanges or lugs, which latter are weak and unreliable, and especially exposed to corrosive influences, the difference between the safe tensional and compressive strain of cast iron also being disadvantageous. It has been suggested that the bolts should be made of steel in order to prevent the severe corrosion, but as corrosion of wrought iron and steel is about the same, it is difficult to see that any advantage would accrue, and, moreover, the different electrochemical properties of the metals may set up galvanic action and accelerate decomposition, instead of diminishing it. On the Bombay, Baroda and Central India Railway, some of the external bolts, 1¼ inch in diameter, all of which were above the water level, were found to be so corroded, after only four years' exposure, as to be useless. Internal flanges were used below the water level. When the lateral strength of a structure depends upon the secure bolting of the flanges of the piles, which, in this instance, were 30 inches in diameter and 1 inch in thickness, the importance of their always being unimpaired is evident. Plates and flanges can be evenly and equally painted, but it is difficult to protect bolts over all their surfaces, and dipping in approved tar-asphaltum composition, when·they are heated, is probably the best way to preserve them from corrosion. It is the want of equal protection which produces local corrosion, and causes them to be vulnerable, or sound in one place, and corroded in another. It has been proved, by experience, that when brass bolts have been used to fasten the planking or wooden sheathing of ships to the steel or iron plates of a vessel, no deterioration has resulted, whereas iron bolts in the same situation have soon become corroded.

If lugs on the piles and flanges must be used for holding the tie-bolts, the holes in them should be drilled, so as to ensure that they are truly cylindrical and give a direct and equal bearing. In cooling it is sometimes found that the joint flanges break away from the lugs, which may consequently be only attached to the column, thereby failing to support and strut the flanges. Lugs and such projections are, in addition to being a very weak arrangement, decided aids to corrosion, and it is much better if the ties pass through a column by some water-tight device or around it by means of straps or rings passing round collars. Cast-iron lugs on columns are particularly liable to failure from a sudden strain or shock, for in a casting it is usually found that any considerable projection will have impurities in it, and that at the point where the projection or lug joins the column, the metal will be spongy and porous; corrosion will, therefore, be especially severe wherever there are lugs and bolts, &c. The edges or outer sides of the lugs are sometimes unduly subject to stress if the pins or bolts have not sufficient bearing area, do not fit properly, and are unequally strained ; then corrosion will complete the failure of the joint or the bracing.   There can be no question that passing bracing tie-bolts through the columns, as at the Huelva Pier, designed by Sir G. B. Bruce, is a far better system than placing them through eyes in lugs cast on a column, for the lugs may be wrenched off.   Care, however, should be taken that no air or moisture is thereby admitted into the interior of the column through the holes made in it for the tie-bolts, or corrosion will occur.   This method, or that of passing the bracing round a pile or column between two collars, is an excellent arrangement, as a reliable union with the piles is obtained, the parts are rigid, and local corrosion much lessened.

There is reason to believe that bolts of different diameters have not the same strength per unit of section.   Mr. Brunel found, in testing some iron bolts, all made of the same iron, that a $1\frac{1}{4}$-inch bolt broke at 23 tons per square inch, a 1-inch at 25 tons per square inch, and a $\frac{3}{4}$ and $\frac{5}{8}$-inch at 27 and 32 tons per square inch respectively.   Hence, corrosion in bolts of the larger sections may be of more relative importance than in those of smaller dimensions.   Usually, when made from the same metal, square or round bars are stronger than plates by at least 3 tons for every square inch of section. Tie-bolts are sometimes used either to tie a fender pile to a wall, or to bind together a pile and a wall.   The nuts and washers of such bolts should be designed so as to specially protect them from corrosion at the front and the back of the wall, for the tie-rods may be in a good state of preservation, being surrounded by the material of the wall, and their nuts, washers, heads, &c., be much corroded.   If the heads are below the surface, and the recess be completely filled with Portland cement, they will be preserved from corrosive influences.   See sketch of bolt, Figs. 1 and 2, in Chapter XII.   In certain sea water, iron bolts are not so durable as wooden treenails, if made of exceptionally durable timber, as the blackwood of Australia, teak, &c.

In all metallic hollow pile structures, but especially those for bridge piers, promenade or landing piers, and similarly submerged or occasionally submerged works, in order to protect them from corrosion, it has to be determined whether the interior of the piles shall be filled with some preservative substance, be left open, or the surface be coated. With regard to filling the interior of a hollow pile, the chief object is to prevent water accumulating in the column, and fresh supplies of air, especially moist air or steam, and all corrosive agents having contact with the metallic surface. Metallic cylinders or columns that are not filled with hearting, should be so constructed that they can be thoroughly examined and painted ; this, of course, can only be done with those of considerable size, and ventilation by means of man-holes and air spaces should also be provided, and means adopted by which any moisture confined in a filled-in iron cylinder or column will be led away, and not allowed to accumulate. Every care should be taken, in filling columns with Portland-cement concrete, that the unequal contraction of cast iron and concrete by cold is provided against, also that a non-swelling concrete is used, or internal strains will occur, and may cause a column to crack. Weeping holes can be provided in order to allow of the escape of water accumulated in the pile, or proceeding from an excess of moisture in the hearting, so that any water, after the concrete has set, can flow away. Cracking and bursting of a pile during frosty weather will then not occur. Apart from the expansion of the internal water during freezing, the contraction of the iron during cold weather upon a firm and compact hearting may be injurious, for if concrete filling is used, and it is solidly rammed, the contraction of the iron ring will compress it, or a column will crack or burst, but it is an excellent preservative against corrosion. Columns have been filled with concrete rammed tightly to protect them from rust, and to avoid painting, but unless provision is made in piles of the usual dimensions employed in piers for the contraction of the iron ring, although the column may not crack or burst, some strain must be brought upon the iron from this cause. Wooden staves, wedges, and planking have been fixed between the iron and the concrete in cylinder-bridge piers, but tarred felt is to be preferred to relieve the metal from this strain. In piles of small diameter, it is unimportant. Hollow piles are sometimes filled with clean dry sand. Before any filling is deposited in a pile, its interior should be thoroughly cleared of all dirt and débris. If clean dry sand is available, the interior of the piles may be filled with it, as it has been found that water was so excluded, but, of course, impervious Portland-cement concrete is to be preferred. If concrete be used, no gravel should be mixed with it, but clean sharp sand; the Portland cement should be non-swelling, quick drying, and only just enough water be employed in mixing to enable it to properly combine with the cement, so that no moisture remains in the column. The concrete should also be rich in Portland cement, so as to prevent any water permeating the mass that may from any cause penetrate into the column, for a weak cement concrete is porous, there being capillary orifices in it, and gases can also pass through it, but Portland-cement concrete is an excellent protection against corrosion if properly proportioned and applied.*   To prevent corrosion in the interior of metallic piles or columns in which the hearting does *not* support the load, strength is not required, but the concrete should dry and set quickly and equably, be thoroughly homogeneous and impermeable, and be able to resist any corrosive influences to which it may be subject. All piles, columns, and pillars used in a building, or of too small dimensions to admit of inspection, can, after being properly cleaned and coated, be sealed at the top and bottom to prevent the admission of air or moisture.

The hollow cast-iron piles of the Morecambe Bay Viaduct, erected in 1853, were boiled in a preparation of coal-tar and asphalt for half an hour, and the interior was simply filled with

* For full information with regard to concrete, see ' Notes on Concrete and Works in Concrete,' second edition, published by Messrs. 'Spon, Strand, London.

clean sand.   Mr. Brunlees, the engineer of the viaduct, found that water was thus excluded, which is necessary, or the piles would probably burst in frosty weather.

In screwing hollow cast-iron screw piles having *open* ends, not only to lessen the chance of breakage, but also to retard corrosion, all earth in the pile should be removed by a shell-auger as it proceeds, but *not* in advance of the bottom.   Such a pile, as it has open ends, should be filled with water-tight Portland-cement concrete, and not sand, so as to exclude the atmosphere and damp.   In most soils it is preferable to have a gimlet-pointed screw pile if the diameter is not too large, as then the interior is less liable to corrosive influences.

Examination has shown that in iron pile structures in ordinary sea or brackish water, and, in a lesser degree, in tidal river water, such as the Thames, corrosion appears to be greater in cast-iron piles at or about low-water mark than above it, and does not apparently energetically extend much below that level.

CHAPTER XIV.

NOTES ON CORROSION IN BRIDGE FLOORS, ROADWAYS, AND PLATFORMS.

With regard to bridge floors, roadways, and platforms, and corrosion: strength, durability, lightness, and incombustibility are the chief points to be considered. In order to lessen the dynamic effect of a rolling load upon a bridge, elastic support between the rail and the girder should be provided, for if a longitudinal beam merely rests upon the upper flange of a girder, and the rail upon such timber, the dynamic effect is increased, and is considerable. The tendency in wooden platforms for railway bridges will probably be to increase their thickness under the rails. Their renewal will therefore be more expensive, and inspection of the girders, cross-girders, and floors not so easy; and it will be the more important to provide against the effects of corrosion.

Flooring plates are sometimes placed on the bottom flanges of the cross-girders, and the permanent way is the same as that on the formation of cuttings and embankments, and is unconnected with the platform; the ballast being purposely of extra thickness, even as much as 18 inches, so as to counteract the momentum of a train being suddenly brought to rest, and the transmission of such a thrust direct to the girders, piers, or abutments. A rigid floor, such as Portland-cement concrete on a metal trough, cannot be an elastic medium and therefore any special provision against corrosion may require to be modified, in order to construct the floor so as to neutralise or lessen such strain. A water-tight floor to an ordinarily-constructed bridge is not easily

obtained. In many cases zinc protectors have been fixed
upon the under side of girders over footpaths, to keep the
pavements and footpaths dry, as it is practically impossible
to prevent water percolating through ballast and joints of
timber. In many floors, even if ample provision is made for
leading away the surface water, it will find the course of
least resistance, and this may not be towards a drain ; consequently, in designing a floor, this should be considered, or
corrosion may be considerable, and unequal in intensity.
Corrugated steel and wrought-iron floor plates can be made
nearly water-tight, but provision against corrosion is required; It is not easy, however, to make the whole of a
bridge floor constructed of built-up troughing permanently
water-tight, unless it is covered with asphalt, or other
similar coating of an anti-corrosive and impervious nature,
for all metallic floor plates are subject to contraction and
expansion, consequent upon changes of temperature, and
therefore the joints are likely in time to admit moisture.

The angle or T-irons to which the floor plates are attached should have sharp, clearly-defined angles, so that the
plate edges can very closely rest against them, and not have
round angles, as then they cannot be made to fit so well,
and each will form a ready receptacle for active corrosive
influences. It is necessary to exercise considerable care in
order to make the edges of the floor plates perfectly flush,
and to tightly fit the T-iron bearers, and make them rest
evenly. If this is not done, rapid corrosion of the rivets,
T-irons, and ends of the floor plates will almost certainly
follow. Everything should fit perfectly, and the whole floor
be as one mass. Bends and angles, unless effectively and
permanently covered, form receptacles for water and substances inducing corrosion. Many railway and road bridges
have floor plates of corrugated wrought iron, and consist of
a series of troughs. If any moisture or dirt accumulates in
the depressions of the bends, actively corrosive conditions
will be produced, therefore their surface should be well
protected before erection, and, if practicable, a non-corrosive,

impermeable material be laid upon them. If water be held
by a trough, the side plates, unless completely protected by
an anti-corrosive coating, may quickly corrode, and the rust
would probably gradually fall off, sink to the bottom, and
form a skin or deposit of a brownish hue.

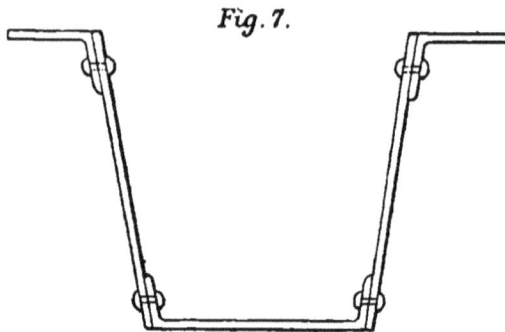

Fig. 7.

Flooring plates are sometimes joined together as shown
in Fig. 7, which cannot be said to be a good arrangement to
prevent corrosion, as there are two top and two lower joints,
the rivets can work loose, and water may permeate between
the plates presumed to be always in close contact, and then
corrosion will occur, and unequally. Fig. 8 is a preferable
joint, for there can be but little lodgment of water on the

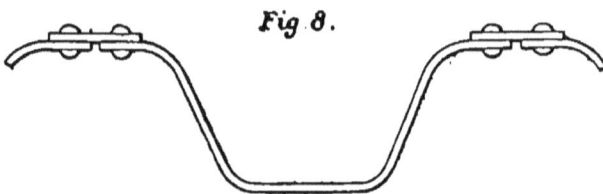

Fig 8.

joint plate, and there is no joint at the bottom of the trough,
and, if the rivets should fail from corrosion, the trough will
probably not buckle seriously; but if any of the rivets in
Fig. 7 become corroded, and the plates act as single pieces,
the strength is impaired and may be seriously so. In any

case, Fig. 8 is a much better arrangement than Fig. 7, and is more easily examined.

In preference to trough plates—which, unless completely filled with Portland-cement concrete or other water-tight material, are liable to hold water, or cause moisture to reach the plates—in Fig. 9 an arrangement, which appeared in 'The Railway Engineer,' is shown for places where the headway is of importance, as the distance from the underside of the girders to the top of the floor plate can

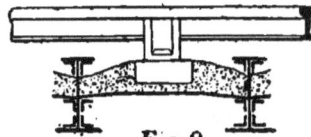

Fig. 9.

be but a few inches, and the ballast be laid upon it, and be as little as 6 inches in depth, so that to inspect the plate, or to repaint it with tar-asphaltum, or other anti-corrosive composition, the ballast has but to be removed, and the sleeper supported. Any corrosion can thus be prevented. In the case of trough plates a greater depth of ballast must be used, perhaps twice as much, and deadweight is, therefore, added to a bridge for about half the extra depth, because there is no ballast on the upper portion of the trough. The junctions of floor plates or troughs with the main girders are generally likely to leak after a time, and should be carefully inspected periodically. Obviously, flooring plates which have the fewest joints are the least subject to local corrosion. Any local accumulation of water, caused by corrugated plates, stays, or deposit of any kind, should be avoided.

Portland-cement concrete and asphalt can readily be made to be water-tight, but macadam, metalling of any kind, stone and wood paving, unless kept in excellent repair, and placed upon a water-tight Portland-cement concrete bed, properly proportioned, made, and deposited, are not water-tight, and cannot be considered as affording protection from corrosive influences, drainage, or rain water, or the dripping from locomotives, &c. When porous material has to be used for the roadway, a good plan to prevent corrosion is to cover the metal with a layer of asphalt or impervious Portland-

o

cement concrete, or an approved bituminous concrete. Bituminous concrete is sometimes used for the first or bottom
layer, and then Portland-cement concrete, and asphalt at
top. Such a roadway covering should be water-tight if it be
sufficiently elastic, and properly made. Before any asphalt
or other coating is placed upon the floor plates, the whole
should be scraped, swept, and made quite clean, dry, and
free from all decaying or corrosive matter. This is best
done in convenient lengths, so that, after cleaning, the floor
plates and bearers are not exposed to the atmosphere or dirt
longer than is necessary. It can be so arranged that, almost
immediately after being thoroughly cleaned, the plates are
coated. The asphalt can be dressed to the proper falls,
and very small short pipes, or openings, covered with grating
flaps, to carry off rain and surface water, can be inserted. If
it is desirable to prevent leakage from a wooden platform
of a bridge, in order to lessen any corrosion of the girders
beneath, the flooring can be tongued, and the joints caulked,
so as to make it like the deck of a ship. The floors of street
bridges, owing to dung and other corrosive matter, which is
not present on the floors of a railway bridge—other conditions
being of a similar character—cause more active corrosion in
the iron or steel main or cross-girders, or other beams upon
which they may rest. Perhaps the worst floor for corrosion
is planking upon timber beams, as, in addition to any acid
that the wood possesses, there is the almost constant more
or less damp state of such a wooden covering, for wood
when perfectly dried is extremely hygroscopic. There is
also the leakage of the dirty impure water and snow from
the roadway to the bearers and girders. The chief redeem-
ing feature would appear to be the necessity of a timber
floor of a street bridge having to be frequently renewed, and
consequently the girders have to be occasionally inspected.
A wooden block floor on concrete is an improvement on the
old planking on wooden bearers. The trough system, the
metal being well coated with asphalt in every direction,
is, however, much to be preferred, as the filling can be

light, dry, and consist of damp-absorbing material, such as ashes, care being taken to obtain sufficient rigidity for the upper surface of the road, and then the roadway can be made of any desired material, and any part repaired or inspected without structural alterations, as in timber floors.

In any iron or steel structure liable to be covered or be-spattered with street mud, it is advisable to know the nature of the mud to form an idea of its probable corrosive effect. Dr. Letheby (see ' The Chemical News '), in a series of experiments conducted under his direction in the City of London, found that the composition of mud from the stone-paved streets of the city, compared with fresh horse-dung and farmyard dung dried at 300° F., was :—

Constituents.	Fresh horse-dung.	Farmyard dung.	Maximum organic.	Minimum organic.	Average.
Organic matter	82·7	69·9	*58·2	*20·5	47·2
Mineral matter	17·3	30·1	41·8	79·5	52·8
	100·0	100·0	100·0	100·0	100·0

The largest amount of mineral matter is always found in the mud in wet weather, when the abrasion of the stone and iron is the greatest. It then may amount to 79 per cent. of the weight of the dry mud ; whereas in dry weather it does not exceed 42 per cent. Taking the average of all weathers, the amount of horse-dung to abraded matters is 57 per cent. In the case of wood pavement, the amount of organic matter in the dried mud was larger than in the case of the stone pavement. It was 60 per cent. The amount of moisture in the street mud varies to a considerable extent, according to the state of the weather, but it is rarely less than 35·3 per cent. of the weight of the mud in the driest weather, the average in ordinary weather being 48·5 per cent., and in wet weather it ranges from 70 to 90 per cent. The mud was always so finely comminuted that it floated freely away in a stream of water, and the inference is that it would not subside to any great extent in a sewer with a moderate flow of water. This information affords a good criterion of the amount of ammonia and other corrosive agents to be met with in street surface water, also that which percolates into the soil and passes over any metal work before it flows into the sewers.

Chloride of calcium has been used in Rouen for watering the principal streets, and it was found that it impregnates the soil with hygrometric matter, and maintains its humidity for a week, the economy effected being 30 per cent. It is to be remembered that any such solution produces dampness and accelerates corrosion, apart from any special effect of the chemical introduced. Sea water, or water mixed with salt, also acts similarly, but not so powerfully. Where damp sea air reaches the under side, and rain the upper portion of iron or steel bridge flooring, corrosive influences will be active.

The drainage of the platforms of bridges is of importance, and the protection of the floors from the drip of locomotives, and the removal of any dust or earth that may accumulate. All ballast on bridge floors should be carefully selected, be dry, and entirely free from earth, decomposing, or friable matter, and be capable of absorbing water, but only to quickly evaporate it; small shells, and clean washed gravel and sand, being better for this purpose than broken stone, as there is dust with the latter, unless it is carefully washed, and it frequently is not so able to withstand weather influences as the other materials mentioned. Ashes are naturally hygroscopic, and, owing to the calcining of the calcareous matter, which usually is found in fuel, they readily assimilate water by chemical action. The portion of carbon which remains unburnt acts as an oxidising and antiseptic agent in the destruction of the putrefying organic matter occurring in damp earth or marshes, and, being reduced to a fine powder, the ashes form a compact mass, which prevents the filtration of water, and the formation of pools. This was found to be the case in using ashes to reclaim deadly malarial swamps in some parts of the South Austrian Railway, and it is equally effective in any similar situation. By crushing, coke has been reduced in bulk by about 40 per cent., and this may be taken as a rough approximation of the proportion of pores to solid matter.

With regard to asphalt, it is carbonate of lime naturally impregnated with bitumen in very variable proportions, but which for road-making should be limited to between 7 and 12 per cent. Exposed to the atmosphere, asphalt gradually assumes a grey, almost a white tint, caused by the bitumen evaporating from the surface and leaving a film of limestone. Gas tar is decidedly the worst form of bitumen for paving purposes, as it passes from the dry to the liquid sate, and *vice versâ*, according to the season. It is also very brittle. Bituminous coatings are frequently used so as to afford a solid and water-tight layer. Mineral tar, powdered lime, and fine baked sand have been used, but they deteriorate in time. However, by fresh coating with about equal parts of tar and powdered lime, laid on hot, as is sometimes done in reservoirs, they become solid and water-tight, but a neat Portland-cement coating is of a more permanent character, except where considerable elasticity is required in the coating. When asphalt is specified for covering road or railway bridge floor plates, it is important to know that asphalt is being used, and not a mixture of ground limestone, ground slate, and Trinidad bitumen, or ground chalk, fire clay and gas tar, which is dull and black, has a tarry smell, and a hard metallic sound when struck against iron in cold weather. These are sometimes passed off as asphalt. The former is not so good as real asphalt, and is nearly as dear, and the latter becomes soft in summer and cracks in winter, and therefore admits corrosive influences. No preparation of gas tar is asphalt, nor is bitumen, shale grease, pitch from suets, or Stockholm tar. It is, therefore, always advisable to know that the asphalt comes from well-known mines, or that the quality is vouched for by experts. If the layer of asphalt is of a total thickness of $\frac{3}{4}$ inch, it is generally sufficiently thick, and it is usually laid on in two layers. A kind of sawdust asphalt, made with coal tar and sawdust, is sometimes used to cover floor plates, instead of concrete, in order to make a very considerable saving in the weight of the covering. If spurious mixtures called asphalts are used, corrosive influences must be much increased, and the composition may cease to be an antiseptic and anti-corrosive material.

199

CHAPTER XV.

CORROSION IN WATER-PIPES, SEWERS, ETC.CORROSION IN WATER-PIPES, SEWERS, ETC.

WITH regard to the corrosion of cast-iron water-pipes, it varies considerably, according to the locality, composition of the water, &c. In some cases, the pipes appear to be almost free from corrosion; for instance, the cast-iron mains of the Seville Waterworks, laid in 1884, as far as can be ascertained, were free from corrosion and deposit ten years after. Their state, however, in many other cases has become most serious. The pressure of the water in the mains has been used on the Appold system to propel or drive pipe-scraping machines, having steel scrapers, through some water-mains. To show that corrosion does occur to a considerable extent in water-pipes, it has been recorded that the motion of the scraper has been arrested for about an hour until the scrapings accumulated in front of it were washed away. At Grenoble, the pipes became covered with rust carbuncles, which M. Lory, of Grenoble, found consisted mainly of hydrated oxide of iron, and contained from 5 to 10 per cent. of organic matter. It is believed that, apart from the general chemical composition of the water, which obviously will cause different effects as it varies, water containing organic matter in solution quickly attacks iron pipes. That some kinds of water do have a destructive action on iron is shown by examples. The magnitude of the corrosion depends in great measure upon the quality of the iron, see Chapter III. The external corrosion, should the earth be at all saturated with foul water or gases, but not otherwise, is likely to be greater than the internal corrosion; however, if the water is fit for drinking, and does not contain much air in combination, internal corrosion will *not* be severe, provided the iron is of good quality. If bad, weak, heterogeneous, and carelessly made metal is used for pipes, it will soon corrode, and usually in tubercular form. In uncoated pipes, corrosion commences and proceeds somewhat uniformly if the metal is of good and homogeneous quality, and its degree of action gradually increases. It usually presents a rough uneven surface, although it will prevail over it, if the pipes are uncoated. If protected by composition, the surface will not appreciably corrode until the coating becomes exhausted by time, or has weak places in it forming active centres of corrosion, upon which bunches of rust are formed. Therefore, the efficiency of an anti-corrosive coating much depends upon its being thoroughly incorporated before being applied, and evenly spread.

With respect to the deposits which form upon, adhere to, and have been removed from water-pipes, some substances are stones, lead, pieces of pipes, spades, crow-bars, peaty matter, gravel, calcareous earth and slime, pieces of wood and mussels; and these from pipes varying from 6 inches to 24 inches in diameter, and in one case the gain in delivery of the main pipe after scraping was not less than 25 per cent., and averaged 45 per cent., and in special cases very much more, even as high as 250 per cent. Rusty pipes also contain algæ and other vegetable and animal growth. Earthy salts contained in spring water also form a deposit upon the pipes. The nature of the deposit will vary according to the character of the water; in some cases, carbonate of lime may be deposited, in others, hydrous oxide of iron, &c., &c.

Mr. Mansergh, M. Council, Inst. C.E., gives the result of an analysis of iron rust and deposit, made by Dr. Harker, the medical officer, raked out from the 8-inch main pipes of the Lancaster Waterworks, laid in 1852, and scraped in 1878, as the reduced discharge demanded such clearing * :—

* See 'Minutes of Proceedings,' Inst. C.E., vol. lxviii.

Chemical Examination.—Water, 64; organic matter and water of composition, consisting chiefly of brown insoluble vegetable matter, common to all soils, 23; mineral matter after incineration, 13 (consisting of iron protoxide 7, silica in minute crystals, silver sand 2, alumina 2, lime and other soils and loss 2, total 13): total 100. The larger flakes of scale removed measured $\frac{3}{8}$ inch in thickness. The material was of a light spongy brown appearance. The mean result of two examinations is given, as by evaporation the substance changes in density from day to day. The scrapings were fine soil matters deposited under high water-pressure, and iron rust from the pipes.

In considering the probable corrosion, it is well to remember that water always has a tendency to follow along a pipe embedded in the earth, especially when its mouth is entirely submerged, and there is a considerable head of water on it, hence corrosion may be expected. Although pipes may be efficiently laid and coated, and in dry soil, leaks at the joints and through defects in the castings may make the ground in a constantly damp condition, and therefore one favouring corrosion. It is always advisable that each length of pipe should be either properly tested, to discover any concealed flaws, with a pressure in excess of that it will have to sustain, either at the foundry or in the trench, and all defective lengths rejected.

Cast-iron pipes are usually preferred to wrought-iron for durability, and if they are coated with Dr. Angus Smith's composition soon after casting, their durability is increased, but they have rougher internal surfaces, which increase friction and the formation of deposits, and may have hidden flaws. Wrought-iron pipes are sometimes adopted simply to save the cost of carriage; for instance, in order to save the enormous expense of transport, at Kimberley Waterworks, the pipes are of wrought iron, $\frac{1}{4}$ inch in thickness, instead of being ordinary cast-iron pipes. In a dry climate, where the rainfall does not permeate more than about two feet or so, if wrought-iron pipes are laid in such soil as clean sand, the opinion has been expressed that any objection on account of corrosion does not hold. Such may be the case, and can only be proved by time, but if the sand is impregnated with saline or other actively-corrosive substances, and the pipes are not coated by an approved system, it cannot be shown that serious corrosion will not result. Their lightness as compared with cast-iron pipes, and absence from shrinkage-strains in castings—which cause weak places, flaws, air-holes, and variations in thickness—are decided advantages, as also their ability to resist sudden shocks, such as those caused by the shutting of valves or excess of water, but their less thickness is not in their favour, although, on this point it can even be said the cast iron may change in character, and, further, if flaws or air-holes exist, each will be an active corrosive nucleus. Wrought-iron pipes are not often used in Scotland for subterranean petroleum conduits. In many Scotch oil-works, cast-iron pipes are adopted, but, on the other hand, cast-iron for the man-hole mountings of petroleum sheet-iron tanks has been condemned as liable to crack in frosty weather, and also by heat, and it is considered should not be used.

Experienced opinion has declared that a $\frac{1}{4}$ inch in thickness steel pipe is safer than a 1-inch cast-iron pipe under similar conditions, and whereas cast iron will break on undue loads being applied, steel will only flatten or bend. A large steel water-pipe, five feet in diameter, has been recently laid in Paris. The thickness of the mild rolled steel is 0·235 inch, and is made slightly thicker than the stress limit, 3·8 tons per square inch, required, in order to allow for oxidation. The rivets are of mild steel, and are allowed to stand out if the heads are round, but the tubes may be lap-jointed. So far as regards strength, a steel pipe is to be preferred ; however, consequent upon the thinner section of metal, corrosion may be more serious.

In some experiments made on galvanised service-pipes at Boston, it was found by Messrs. Nicholls and Russell that the insoluble precipitate formed by the action of water on zinc, when analysed, consisted of zinc, water, and carbonic acid, forming a zinc hydrocarbonate. Tests were also made of a pipe coated with a composition of lead, tin, and antimony ; and of a brass pipe. Lead and tin in the former, and zinc and copper in the latter, were found in solution.

Carbonic and nitric acids are most injurious to lead. Carbonic acid generally exists in the atmosphere, and it readily attacks lead in the presence of moisture. Lead pipes are liable to be attacked by carbonic acid in a soil, if not protected by being laid in wooden troughs and surrounded with pitch, or otherwise protected. Wrought iron is attacked by any ammonia or sulphur compounds left in gas. Lead is quickly altered by contact with damp plaster. Sulphate of lead is thus formed, which is a most energetic oxidising agent. Saltpetre, which may be found in the steining of wells and damp cellars, is also destructive to lead pipes in pumps. In damp air, a bright or freshly-cut surface of lead soon becomes coated with a thin scale of grey oxide, which adheres closely to the metal and prevents further oxidation. Lime-water, lime-putty, or lime-mortar attacks lead, and forms a pale yellow deposit of oxide of lead in a short time, if the air is not excluded, the presence of air and moisture being necessary for oxidation. If the mortar is decidedly alkaline, the effect will be greater. This action seldom happens in water-service pipes, as the necessary conditions do not then often exist, but it may do so if lead be used for packing in engineering structures, and lime-mortar or limy compounds, or water are present.

A piece of pipe, which for six years had formed the supply pipe to a fountain basin, was analysed at the chemical laboratory at Berlin. It contained 99·05 parts of oxide of lead, the residue consisting of carbonic acid, with traces of lime and silica, which latter, it was considered, may have been due to small quantities of cement adhering to the outer surface of the pipe. The pipe was much corroded where embedded in cement, the corrosion being most marked at the end of the pipe next the basin, and the effect diminishing as the pipe receded from the water. The pipe was coated with a chocolate-coloured layer of oxide of lead of the hardness of glass, which adhered strongly to the metal. Dry Portland and Roman cement were found by Dr. Peschke to have no action upon lead ; the presence of water appears to be neces- sary to effect corrosion, although no explanation, it was con- sidered, could then be given. It has been found in Germany, in taking up some lead pipes, that they had become brittle and porous, and some of the lead had been converted into a basic carbonate white lead, and this had only occurred where the lead pipe had been in contact with mortar or cement. By experiment, Dr. Rossel found that lead buried in most earth that contains chlorides, saltpetre, and sal-ammoniac, lost weight, but to a much less degree than in mortar ; and that the sulphates, like plaster of Paris and Glauber salts, had no action upon lead—however, this does not quite agree with the experience of some others—neither had the carbonates, like chalk, soda and potash, nor the silicates, sand and clay. He considered lead pipes should never be brought in contact with any kind of mortar or cement. Mr. Reichardt, of Jena, examined, a few years ago, the leaden pipes which had been in use for upwards of 300 years for the supply of water from springs to the town of Andernach, and found them to be coated internally with a layer, about $\frac{1}{2}$ millimetre in thick- ness, of phosphate and chloride of lead, with a little free oxide of lead. Very small quantities of lime and magnesia were present in this coating, and the metal of the pipes, after being in use for this long period, was perfectly good in quality. Various experiments were made, and showed that it was important the lead pipes should always be full of water, so as to exclude the air and prevent oxidation of the lead, and not alternately emptied and filled, as, if oxidation occurred, the metal readily became soluble in water.

To prevent or lessen corrosion on the outer surface of water or gas-mains, perhaps there is no more effectual plan than surrounding them with Portland-cement concrete, properly proportioned, and mixed so as to be impervious, continuous, uniform, and an unvarying protection.* In special cases, where severe corrosion is expected, and iron pipes must be used, they can be surrounded with Portland-cement concrete, after receiving a coating of tar-asphaltum composition, and on the inside be covered with a 1 of Portland cement and 1 of washed sand coating, about ¾ inch in thickness, and so manipulated as to present a smooth, glassy, and water-tight surface. The cast-iron circular culverts, 8 feet in diameter, in the Langton Dock, Liverpool, Mr. Lyster † has lined with a ¾-inch coat of Portland cement mixed with fine gravel, kept in place by dove-tail ribs cast at near intervals on the inner surface of the pipes (see Fig. 10).

Fig. 10.

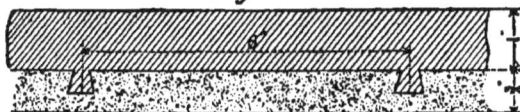

Periodical examination has shown the condition of the pipes continued to be perfect. This system can be readily applied to pipes of large diameter, and could, by an arrangement of an annular movable board, to hold the concrete until set, be adapted for any reasonably small diameter. The conduit of the Paris new water supply consists of iron circular pipes about 5 feet 8 inches in diameter, having an outside lining of masonry 8 inches in thickness, and an inside coating of ¾-inch cement over the lower two-thirds. Where the pipes are under pressure, they are lined inside all round. It has been found, in repairing and replacing water-mains embedded in concrete and used for the Vienna water supply, that the oxidised iron entered into an intimate combination with the concrete.

If the under side of an iron or steel water-tank, for some reason, cannot be made to be accessible for inspection, it

* See 'Notes on Concrete and Works in Concrete,' published by Messrs. Spon, Strand, London.
† See 'Minutes of Proceedings,' Inst. C.E., vol. c.

should be closely and firmly bedded in impermeable Portland-
cement concrete. An allowance for corrosion should always
be made in determining the thickness of the plates. At
Bordeaux, an iron service reservoir, in two basins, each
55 feet in width, 130 feet in length, and 6 feet 7 inches in
depth, with semicircular ends, was constructed on columns
resting upon bed stones on concrete, placed 13 feet apart
longitudinally, and 18 feet 6 inches transversely, because of
the very bad foundation, and the necessity of placing the
bottom level of the reservoir a few feet above the ground.
In order to provide against corrosion, the bottom is everywhere made accessible for painting.

Tie rods and stay bolts are used in cast-iron water-tanks,
and, as they generally corrode quickly at the points of
attachment to the tank plates, the strains on the tank become very severe, and are local. Cast iron is not a proper
material to employ, for such a tank may explode, but wrought
iron or mild steel plates should be used, as they would bend
before giving way. There need be no fear of corrosion in
wrought-iron or steel tanks if they are properly coated with
tar-asphaltum or other really anti-corrosive composition.
Unprotected cast-iron plates in the beds of some waterworks
have been found to last but a short time, because of the action
of the water upon them.

The methods of making joints for metallic pipes are very
numerous. In order to prevent corrosion, special care is necessary to secure a tight joint. The steam-pipes of the New
York Steam Distribution Company have faced joints, and
are rendered steam-tight by the introduction of a gasket of
thin corrugated copper. The gaskets are painted with thin
red lead, thus excluding moisture. The corrugated copper
gaskets were so made that the outer diameter cleared the
insides of the flange bolts, and were dropped into the pipe
joint from the top after putting in place two or three of the
lower bolts to centre them. This much facilitated the work,
and no difficulty from leakage has ever occurred at such
joints, notwithstanding the very considerable erosive action

of steam jets. As an additional precaution, the gaskets were
painted with thin red lead, as previously mentioned. Red
lead putty joints were first used, but they took a considerable
time to complete, and portions of putty were squeezed into
the pipe. A recently-adopted method of making a joint
for pipes is by carefully cleaning and painting the joints,
before the pipes are laid, with thin lead paint, no lead being
used. Tape and red lead, and millboard and white zinc have
also been employed, and special proprietary compositions of
various kinds. In the case of sewer-pipes, cement, tarred
gasket, and other water-tight joints are used. Joints of pipes
are also sometimes well caulked with tarred gasket, the
remainder of the joint being run with molten lead 2 inches
deep, and properly caulked afterwards. Sewer joints are
also made good with canvas and red lead. At Croydon, only
lead was used for making the joints in some pipes, the old
lead and gasket joints being abandoned, and the sockets made
very much shorter and smaller than formerly. It is to be
remembered that galvanic action may occur when iron or
steel is joined by lead, and corrosion be thus accelerated.

It is advisable to externally and internally inspect all
metal pipes *before* they are coated with any preservative
composition, because defects of casting are hidden by paint
or tar. Dr. Angus Smith's composition appears to be universally approved when properly applied. Water-pipes should
be treated by being completely dipped in a bath of the
varnish, so as to coat the entire surfaces inside and out, as
directed by him : i.e., the pipes are to be cleaned, and at a
temperature sufficient to expel as much as possible the air
from the skin of the pipe, 600° F. being sometimes adopted,
and to dry the varnish properly, without burning or deteriorating it, but to produce a smooth, even and regular
coating. Common gas tar will not be effectual, as it does
not combine with the iron or harden, but washes off, and the
tar-asphaltum process must be used. Seven to nine years
may elapse before there is any appreciable deteriorating effect
upon a water-pipe coated with asphalt. If tar only is used,

the pipes should be dry before being laid ; if they are wet, not only will the anti-corrosive coating be more or less a failure, but any water will be polluted, and if they are gas-pipes the moist tar will have a deleterious 'effect on the illuminating power of the gas. If a pipe be coated with the most certain preservative against corrosion, and its surface is afterwards damaged or disturbed, a vulnerable place will be caused, and corrosion will follow. By the coating being thoroughly hard before the pipes are laid, the risk of damage is much reduced. Before being connected, all iron pipes should be thoroughly coated externally and internally with Dr. Angus Smith's composition or coal-tar pitch, applied hot by immersion, or with some equivalent satisfactory substance, and the large mains can receive an internal coating of Portland cement, as previously described. The American Waterworks Association, Philadelphia, 1891, in the report of the committee on specifications for cast-iron water-pipes, recommended, " That pipes are to be thoroughly cleaned without the use of acid, heated to 300° F., and plunged into the coal-pitch varnish. When removed, the coating to fume freely, and set hard within an hour."

Pipes not coated before being laid are generally soon encrusted with rust, and deposits, such as carbonate of lime, &c., depending upon the matter in suspension, the character of the water, and other circumstances. It has been recom-mended that all coatings be put on immediately after the pipes are cast, and that the pipes should be tested with oil, and not by water pressure. The great point in cast-iron pipes is not to disturb the skin, and to preserve it at once by coating it. If the metal be chipped or exposed by any other means under the skin, corrosion will occur, and may work its way under the coating or varnish of the adjacent surfaces.

After about four years' use, it was found that the inte-rior of the cast-iron pipe of the Fleet Sewer crossing London Metropolitan Railway, which was coated with red lead, had suffered no deterioration from the constant flow of sewage matter through it. The paint still adhered to the sides, although it was washed away from the invert. The steel submerged sewer, $\frac{3}{8}$ inch in thickness, that is inserted below the bed of the Shirley Gut, which is tidal, Boston, U.S.A.,* is lined with three rings of brickwork in Portland cement, mixed in the proportion of 1 of cement to 2 of coarse, clean, sharp sand, in order to provide against corrosion. The diameter inside the brickwork is 6 feet 4 inches. At Yarmouth, the sewerage syphon-pipes are made of sheet iron, galvanised so as to resist corrosion.

All metallic pipes for conveying sewage should have a thick, hard, and tenacious coating, applied while the pipe is hot, if neither a Portland-cement concrete, nor brickwork in cement lining, is applied, and they are not galvanised; and only metal made to a properly-drawn specification should be used.

* See 'Engineering,' March 16, 1894.

CHAPTER XVI.

CORROSION BY CONTACT WITH WOOD, AND IN GIRDER-SEATS AND BEDS.

In works of construction, metals and woods have very often to be placed in contact, or used in combination, therefore any decay of the timber will affect the iron or steel. It is advisable to consider the effect timber may have upon the metal, and some means of preserving the wood from decomposition.

> " Thro' trunks of trees fermenting sap proceeds,
> To feed and tinge the living boughs it feeds."

It is generally agreed that the decay of timber is due to the sap, which, originally liquid, becomes gradually solid, and then ferments under the influence of the atmosphere, the process being greatly assisted by heat and damp. The air and moisture that are present in the cells of the wood require to be replaced by an approved antiseptic solution, creosote being generally considered the best. The chief points are, therefore, to extract all the sap, make the wood perfectly dry, and then impregnate it with an approved antiseptic, to prevent'fermentation; or, as a makeshift, if this cannot be done, by coating the timber *after* the sap is extracted, and the wood thoroughly dried.

In engineering structures, it is seldom air and moisture can be excluded, hence the decay of wood will proceed and promote corrosion. Wood treated with an antiseptic which has no deteriorating influence on iron will have a decidedly less corrosive effect on iron or steel than ordinary timber, even if it be dried before being used. The decomposition and fermentation of the sap, when wood is exposed to the action of warm, damp air, will continue unless the sap is extracted or the wood properly creosoted or treated, and any fungoid growth on the timber will have an active corrosive influence on metals. There are two conditions especially favourable to decomposition in wood : the confinement of water in it by paint, or a covering producing " wet rot ; " and " dry rot," in which the wood is exposed to warm air. When the deleterious effect of air on wood is extreme, dry rot, which occurs only in timber that is dead, takes place, the residue being brown or black ; and when water, wet rot, which may have commenced before the tree was felled, and the residue of which is of whitish hue. The latter is quicker in effect than dry rot. If the fibre of wood is bruised, it aids the dry-rot fungus. If timber is placed under water, and is not subject to any action of air, wood docs not decay for a very long time, perhaps centuries, if it was sound and free from decaying agents when submerged, although its strength as regards strain may be decreased. By air and moisture wood becomes superficially changed, becoming of a darker colour, the process of oxidation and combustion is slow, simultaneous action of air and water being necessary for rapidity of change.

Timber telegraph-posts decay most at the ground line, because of the incessant changes of moisture and temperature. Some pine foundation-piles in clay, constantly wet, at Albany, N.Y., were found to be not materially decayed after fifty years, but, wherever the moisture was drawn off, the wood did not last more than twelve years.

Wood that is free from large and loose knots, wind shakes, sap, brashwood, worm-holes, is close-grained and even fibred, and which possesses strength, stiffness, hardness, and durability, is the less likely to become decayed, or enable depressed or damp spots to be formed in it. A chemical analysis, however, is necessary in order to ascertain whether it contains anything that will help or cause corrosion of any

metal in contact with it. Wood absorbs water, and, therefore, a girder placed upon it will always be in contact with
a substance that may cause corrosion. The absorptive power
varies very considerably. Some experiments have shown
freshly-cut wood, felled about March, to contain by weight
water to the extent of $\frac{1}{5}$ to $\frac{1}{2}$ of its total weight. Ash contained about 28 per cent., birch 31, English oak 35, fir 37,
pine and beech 40, elm 44·5, red deal 45, larch 48, Lombardy poplar 48, white poplar 50·6, and black poplar 52 per
cent. A large quantity evaporates, but on the hygroscopic
state of the atmosphere and that of the wood approaching
equality, drying takes place very slowly, and is never
complete under such conditions. Air-dried timber usually
contains 15 to 20 per cent. of water.*

Timber imported from warm climates should be examined
as soon as possible after it is unloaded, in order that it may
be known whether it contains any water possessing a decomposing or putrefactive power ; because, if so, although
not apparently so, it is on shipment in a state of decay, and
the damp, warm air of the hold of a ship will help its development, hence the value of thorough drying and expulsion of
air, and impregnation with creosote according to the most
approved manner.

Although tannic acid in wood acts deleteriously on iron
and steel, it is considered to have a preservative effect on
wood ; for instance, the Quebracho Colorado wood of North
Argentina, which is of blood-red colour, is exceptionally
durable and proof against insects, and this is attributable
to the large proportion of tannin it contains, said to be from
15 to 20 per cent. of its weight, which is about 78 lbs. a
cubic foot.

Uncoated or bare iron quickly decays wood in contact
with it. In an able article in 'Engineering,' August 20,
1875, the destructive action of iron on wood is so clearly
stated that it is desirable to make the following quotation

* See 'Bulletin Mensuel de la Société des anciens Élèves des Écoles
nationales d'Arts et Métiers,' vol. x.

from it. "In a damp medium, and exposed to the air, iron rusts and forms sesquioxide of iron. When in contact with organic matters, the sesquioxide yields its oxygen to the hydrogen of the wood, and passes into the condition of protoxide. But this combination rapidly absorbs oxygen from the air, and produces a new oxidation in the wood. The sesquioxide, under these conditions, is really a reservoir of oxygen, which, on one hand, discharges itself at the expense of the wood, and, on the other, fills itself at the expense of the air, and the wood is gradually consumed. Under such conditions, sulphate of copper is more destructive than useful, because, as soon as it is in contact with the iron, it becomes transferred into a salt of iron, which is of itself very destructive."

The tannin and the wood-gum or resin in wood, some recent experiments have shown, are not attacked by the dry-rot fungus, but the coniferin and cellulose are absorbed. In Alsace, it is customary to specify that only raft timber, i.e., that floated down the river, shall be employed, as it has been found timber thus immersed is no longer liable to the attack of dry rot. The reason is considered to be, that the water slowly dissolves out the albumen and salts, and thus deprives the fungus of the nutriment needful for its development. Experiments, by burying fresh sawdust in damp earth, confirm this, for it rots away in a few years, but sawdust, which has been soaked for some time in water and has been thereby deprived of soluble matters, will remain in the ground, under similar circumstances, apparently unchanged, and only slightly tinged on the exterior with earthy matters dissolved from the soil.

It should be borne in mind that if timber is painted when it is sappy, or before it has been properly seasoned or dried, fermentation and decay will be more active. The sap of the wood contains easily decomposed substances, such as albumen, gelatine, gum, &c., and when warm, damp air, is present to aid fermentation, the woody fibres soon become impaired. To winter-felled, seasoned, sound, and air-dried timber, paint

is a protection, but to sappy wood it is not equally so, nor to
timber not thoroughly dry, but effective air-drying has been
known to require four to six years, and therefore an
artificial drying process has to be employed. Paint prevents
moisture penetrating to act as a reagent to decompose the
albumen of the wood. Planed and worked surfaces are in
some well-directed works merely oiled two or three times,
which prevents cracking and drawing, and are not painted
when used indoors.

As a rule, it will be found that the baulks forming the
main support of a temporary bridge or timber platform
decay the quickest. Anything which causes condensation
of moisture decays and corrodes the wood and the iron.
Wounds in the surface of timber, caused by fastenings, nails,
spikes, &c., accelerate decay. Iron in rusting has the effect
of decaying vegetable fibre with which it may be in contact,
and this deleterious action, being additional to that of water
and air, corrosion of the iron and decay of the wood result.
Driving spikes and nails, &c., and fixing bolts, removes or
impairs, to some extent, any but the very hardest and most
persistent protecting covering on the iron. The end of a
spike or nail, or driven fastening, will almost always be bare,
also the screwed portion of a bolt. Covering the heads and
around them, of spikes, bolts, nails, &c., *after* driving or
fixing, lessens corrosion, especially if done so effectually as to
prevent the percolation of air, water, and dirt. This can be
done by Portland cement being pressed upon the head, if it
is sunk a little distance below the surface, so as to form a
receptacle for it, see Chapter XII. Bare iron bolts in
contact with wood rust rapidly, and, if with oak or wood
having much acid in it, very quickly. The comparatively
considerable power of resistance of larch to decay by moisture, the toughness of the fibres, its incombustible and incor-
ruptible nature, and its soundness, whether exposed to the
atmosphere or when buried in the earth, in comparison with
many other woods, are most valuable properties. Its uses
are various. In France it has been employed for water-pipes;

in Switzerland for vine-props. Its other usual applications
need not be mentioned. Its durability, strength, and general
resisting powers cause it to be a valuable wood to use in
contact with iron or steel, in order to lessen corrosion in
places where timber and iron must be in contact, rest one on
the other, or be used in combination. Larch sleepers with
the bark on, it is found, deteriorate more rapidly than when
it is removed, which is believed to be due to the absorption
of moisture in the bark.

Wood having holes in it will have a zone of rottenness
round the holes, which quickly spreads. If the fibre of wood
is bruised, it aids the dry-rot fungus, consequently there
should be as few iron or steel bolts in wood as possible, for
timber united by bolts, whether under the surface of the
earth or above it, is liable to active corrosive influences. It
is found in wooden ships that treenails that have been
squeezed in driving, become rotten quickly. They should be
properly creosoted to counteract this. As holes made by the
Teredo navalis in timber piles cause water and corrosive
influences to penetrate to any iron bolts in them, it is well
to remember that the *Teredo navalis* does not eat *below* the
ground line, nor do they like sewer or gaswork water. The
Teredo navalis perforates wood by mechanical means. Three
circumstances are known to favour its ravages—a moderate
rainfall, increased saltness of the water, and an increase of
temperature. In some experiments made by the Academy
of Sciences of Amsterdam, it was found that all exterior
applications failed. A coat of mail, consisting of nails, was
not only costly, but was not successful, for the covering of
iron and rust did not arrest the ravages of the teredo.
Creosoting, carefully effected, and caused to penetrate the
whole of the timber, and not merely superficially, with oil of
good quality, by thorough impregnation, was alone successful
of all impregnations tried. Sheets of iron, copper, or zinc were
effectual so long as they remained intact and undamaged.
Neither the *Teredo navalis* nor the *Limnoria terebrans* will
attack wood properly and wholly impregnated with good

creosote oil. Piles have been protected in special cases, from a little below the ground line, by clay pipes filled in tightly with Portland-cement concrete. From many reliable experiments in northern seas it appears, if timber be thoroughly dried till every particle of moisture is evaporated before creosoting, and as much creosote be absorbed by the pores of the wood as they can be made to contain, that, when treated in the most approved manner, it is an effectual preservative against the teredo. The creosoting process of preserving the timbers of the bridges, and also the piles of the bridges crossing the numerous bays and inlets, against the ravages of the teredo was adopted on the New Orleans and Mobile Railroad some twenty years ago, as decay of the untreated timber took place very rapidly in the long warm seasons of that latitude. It was most successful, and it was found that there was no danger of fire from the impregnated timber as compared with untreated wood, but the reverse. To show the varying character and increased tendency to cause fouling and decay, and the greater virulence of the attack of marine worms in warmer climates than Great Britain, it may be mentioned that the recent (1885) report of the Committee of the American Society of Civil Engineers mentioned that, although from 10 to 12 lbs. of creosote per cubic foot of timber is sufficient as a protection in English harbours, for the higher temperatures of the sea water on the French coast, and in the southern harbours in the United States, about 19 lbs. per cubic foot are required to secure immunity from attack. A paper read before the Society of Engineers by Mr. John Blackbourn, C.E., Public Works Department, Melbourne, contains much information on ' The action of marine worms and the remedies applied in the harbour of San Francisco, California.' *

The chief object of processes for preserving timber from decay is to completely seal the pores of the wood, i.e., the minute holes through which the sap circulates, with the

* See 'The Engineer,' October 23, 1874.

preservative matter injected into or formed within the pores themselves. It is advisable, whenever timber, preserved by antiseptic treatment, has to be in contact with iron or steel, to consider whether such treatment of the timber may not accelerate corrosion. Most probably it will not, but it can hardly be taken for granted, for treatment by tar process introduces tar acids, and experiments have shown that tar acids absorb moisture before finally evaporating, and it is sometimes specified that as much as 10 per cent. of tar acids is required, in others about 5 per cent., but it may be much less, even as low as 1 per cent.; however, creosote obtained from coal tar contains tar acids to an appreciable extent. In some tests, lasting several years, by Mr. C. Coisne, for the Belgian Government, the creosote oil which had no tar acids in it gave the best results. Some were tested with 15, 8, and 7 per cent. of tar acids. Experiments have shown that wood absorbs moisture when treated with oils containing tar acids, which do not, therefore, conduce to arrest the corrosion of iron or steel.

In the case of timber, metallic salts cannot be as effectual as such known antiseptics as those of the nature of carbolic acid and tar, which coagulate the albumen of the wood. Oil of tar, called creosote, is one of the products obtained from the distillation of coal tar, by which its ammoniacal liquor and other impurities are expelled, and possesses powerful antiseptic properties, hence the superiority of the creosote-oil process of preserving timber. Some of the other processes are Burnetising, chloride of zinc being used; Kyanising, corrosive sublimate being employed; Boucherising, treatment by sulphate of copper. They retard decay for a time, and then usually become redissolved and washed out by rain. Among recent processes introduced are those of impregnating the wood, previously dried, with boracic acid. Soaking timber in naphthalin: this is applied to either green wood or seasoned timber. Also that of Col. Haskin, U.S.A., by treatment of timber in its green state. It thus saves the trouble, time, and expense of drying the wood; but whether it will be effective as a preservative, time will determine, and whether wood so treated has greater strength or not as compared with ordinary green timber, which has considerably less strength than seasoned wood.

The process is simple, for after the wood in its green state is placed in an air-tight chamber and subject to an air pressure of from 150 to 200 lbs. to the square inch, according to the character of the wood, the air being first dried and heated by passing through pipes over a stove, the ultimate temperature, it is stated, varies from 200° to 450° F. All fluid matter is said to be thus caused to be insoluble and coagulated in the pores, and so prevents decomposition; however, nearly 40,000 tests of timber made at the laboratory of the Washington University at St. Louis, under the direction of the forestry division of the Department of Agriculture, have shown that seasoned timber is about twice as strong as green timber, but well-seasoned timber loses its strength with the absorption of moisture. These indicate that green timber should not be used, and, when dried, that moisture should be excluded from it; however, if, by treatment, green timber can be made as strong and durable as thoroughly sound and seasoned timber, much will be gained.

Impregnating timber with creosote, after thorough drying of the wood and expulsion of sap, air, and moisture, appears to be the most generally approved method of preserving timber. Damp or cold air is found to be injurious to the effective impregnation with creosote.

An examination of best Memel telegraph posts, treated by the zinc chloride process, fixed in different earths, Mr. Preece has stated, showed that after thirteen years' use, 40 per cent. of those in sand had decayed, about 33 per cent. in clay, and about 28 per cent. in chalk. This, of course, does not show that the effect would have been the same in the case of iron or steel posts, but it does show that some soils either have a greater decaying or corrosive effect than others, or permit such corrosive and decaying influences to penetrate them. When creosote was used, only a very few piles were at all decayed, about $\frac{1}{2}$ per cent., and those at the top. Comparative tests were most pronounced in favour of creosoting the piles, and Mr. Preece stated that, after thirty years' experience, he had never seen a properly creosoted pole showing the slightest sign of decay.

Testimony from almost all parts of the world bears witness that creosoting timber, when properly done with heavy creosote oil of attested quality, and not light oils of a volatile nature—the heavy oils and not the tar acids being generally considered the desirable preserving agent—is a reliable preservative of wood for engineering structures, always provided the wood was not cut or adzed as in sleepers and piles after creosoting, and the untreated or partly treated timber reached and exposed.

In order to comply with this requirement, which frequently cannot be done in using railway sleepers, piles, bridge-floor timbers, and wharf work, the preservation internally and externally has been obtained by carbolising the interior and creosoting the exterior.

On the Great Western Railway, Mr. W. L. Owen stated that in bridges and viaducts, uncreosoted yellow pine lasts 12 years; creosoted yellow pine, 20 years; creosoted yellow pine in the permanent way, 30 years; timber prepared in any other way in the permanent way, 12 to 15 years; Baltic sleepers, creosoted, 8 to 10 years; ditto, not treated, 5 years. The preservative power of creosoting sleepers being fully acknowledged, on more than one main line railway, where the traffic is exceptionally heavy, it has been a question to decide whether the sleeper will not be worn out by the traffic *before* it has had time to decay. However, *any* decay of the wood loosens the fastenings, and, mechanical action being thereby accelerated, a sleeper's life is reduced when not creosoted, but the extent of the traffic influences the result.

To preserve with tar it is necessary to apply it hot, and it should be heated to boiling point, but not be over-boiled, or the essential oils, which are volatile, will be lost, and they

help the tar to penetrate the wood. Wood tar is considered
by many as superior to coal tar for timber, because it penetrates the wood more easily, and contains a large amount
of antiseptic substances. If wood that will be in contact
with iron cannot be creosoted, it can generally be coated
with pitch oil and coal tar. When iron bolts or rods pass
through a pole or post, it is advisable to char the end that
will be buried, and give the wood above it two coats of
Stockholm tar, put on hot, or Stockholm and coal tar mixed,
or the wood will soon decay and cause corrosion of the ironwork; or it can be blocked off the ground, scorched with
burning shavings, and well tarred. But if nothing can be
done to preserve the wood, decay may be lessened by simply
placing the wood in the ground in the opposite direction to
that in which it grew : the reason is considered to be that
the capillary tubes in the tree are so adjusted as to oppose
the rising of moisture when the wood is inverted.

American experience with cribwork dams shows that a
wooden dam should not be left hollow, but be filled with
gravel, as it tends to preserve the timber, and therefore any
ironwork connected with it, and it is found stone will not
prevent decay of the wood. A committee of the American
Roadmakers' Association recently investigated the subject of
the preservation of timber, and reported that " The theory
that decay is due to atmospheric germs, finds support in the
fact that timber constantly under water does not decay.
That which is constantly dry decays slowly, due, probably, to
the coagulation of the albumen. Timber thoroughly seasoned
by heat decays less rapidly than that treated by any other
mechanical process. This is due to the coagulation of the
albumen. Timber subject to being alternately dry and moist
decays most rapidly, owing, doubtless, to the softening and
drying of the albuminous substances of the timber, thus
rendering it more certain of attack by atmospheric germs."

With regard to girder-seats, it will frequently be found
in girders of *small* span resting on stone or timber, that
corrosion commences about the girder beds. It is always

well in all small and other girders to have a bearing plate riveted to the under side of the bottom flange, so as to preserve the entire sectional area of the bottom flange. A layer or two of tarred felt of the best quality properly placed between any bearing or wall plate is usually effectual in preventing serious corrosion of the girder-bed end, provided no depression is caused in which water can lodge, or moisture, vapours, gases, &c., accumulate. Simply because of such depression, the felt may be very successful in preventing corrosion at one end, and yet not be so at the other. The remedy is to so arrange the girder beds that they can be entirely inspected, and, therefore, to slightly raise them above the level of a pier or abutment. This permits of repairs being more easily executed, and even the ends of girders being sufficiently raised for inspection without interfering with the traffic, or altering the masonry, brickwork, or concrete upon which they may rest. As an additional precaution, the felt can be soaked and bedded in Portland-cement grout.

If timber has to be used for a girder bed, it should be chiefly selected for durability and absence of any deleterious effect on the metal, and especially if it be durable when alternately wet and dry. Wooden wall plates, unless well creosoted, will change in their corrosive effect on metal resting upon them according to the weather, and apart from any acids or other corrosive substances present in them. The ends of girders also sometimes rest upon a pair of planed cast-iron rubbing plates, the lower of which is bolted to the masonry, asphalt roofing-felt in several layers being used upon which to bed them. An arrangement sometimes adopted is that the bearing plates are riveted to the under side of the girders, and rest directly upon the abutments, to which the girders are secured by holding-down bolts passing through oval holes, and allowing for a little adjustment or expansion of the girders, the holes being covered by large wrought-iron washers to prevent the access of dirt and other corrosive influences—this arrangement being for spans between 30 and 75 feet. In those up to about 20 feet they are simply placed upon the wall plate. The ends of girders are not unfrequently bedded upon sheet lead, so as to cover any inequalities and cause a perfect bearing. A non-conducting substance should be interposed between the lead and the iron, or galvanic action may be caused.* A modern specification thus refers to a girder seat on lead : " The girder seats shall be of cast iron, and of the quality specified for cast-iron work. The upper and under bearing surfaces must be planed perfectly true and parallel, and between the wrought-iron girder seats and the masonry must be a sheet or stratum of 12-lb. lead, covered with a sheet of vulcanite, to prevent contact between the iron and the lead." In designing bearing plates, bed plates, rollers, holding-down bolts, frames, &c., to avoid or lessen corrosion, reference to Chapter XII., 'The Influence of Design and Workmanship with regard to Corrosion,' may be advantageous.

In Part II. of this book, following this page, will be found practical information relating to the fouling, &c., of ships, pile, bridge, promenade and landing piers, and similar works of construction, anti-corrosive, and anti-corrosive and anti-fouling paints and compositions, products of great importance in the preservation and protection from corrosion, fouling, and decay of metallic structures, whether submerged or unsubmerged.

* See Chapter VIII., ' Galvanic Action and Corrosion.'

PART II.

THE PREVENTION OF FOULING AND CORROSION
IN SUBMERGED STRUCTURES AND SHIPS.

CHAPTER I.

FOULING BY SCUM, MUD, AND MARINE PLANTS.

In the First Part of this volume the subject is specially
regarded with reference to engineering and other structures
on land. Here, more particularly, attention is directed to
the corrosion and fouling of fixed, floating, and moving
structures in fresh water and in the sea, and to compositions
and paints used to coat their surfaces so as to shield
them from corrosive influences, and to prevent fouling by
marine life of every kind. It is well to define what is
generally understood by the word fouling. It may be said
to mean the adherence of scum, mud, moss, grass, seaweed,
subaquatic plants, mollusca, coral, barnacles, and marine life.
The elements necessary for its propagation are everywhere
present, and have been thus magnificently described by
Milton, London's greatest poet, in ' Paradise Lost ':—

> " And God said, ' Let the waters generate
> Reptile with spawn abundant,' . . .
> . . . .
> ' Be fruitful, multiply, and in the seas,
> And lakes, and running streams, the waters fill : '
> . . . .
> Forthwith the sounds and seas, each creek and bay,

With fry innumerable swarm, and shoals
Of fish, that with their fins and shining scales
Glide under the green wave, in sculls that oft
Bank the mid sea: part single, or with mate,
Graze the sea-weed their pasture, and through groves
Of coral stray, or sporting with quick glance
Show to the sun their wav'd coats dropt with gold;
Or in their pearly shells at ease attend
Moist nutriment, or under rocks their food
In jointed armour watch : on smooth the seal
And bended dolphins play : part, huge of bulk,
Wallowing unwieldy, enormous in their gait,
Tempest the ocean : there Leviathan,
Hugest of living creatures, on the deep
Stretch'd like a promontory, sleeps or swims,
And seems a moving land, and at his gills
Draws in, and at his trunk spouts out a sea.
Meanwhile the tepid caves, and fens, and shores,
Their brood as numerous hatch from the egg that soon
Bursting with kindly rupture forth disclosed
Their callow young."

It is necessary to consider the nature of the substances and life which become attached to a submerged structure, whether fixed or floating, and the means employed to prevent or diminish their adherence.

The formation of scum or mud can hardly be prevented, for it may be figuratively described as the dust of the sea, and, therefore, as practically impossible to avoid as dust on land, which, sooner or later, will become attached to or upon everything that is exposed. However, it may be much lessened by the hardness, polish, freedom from propensity to form attachments, and non-viscidity of the exposed exterior surface ; and also by its possessing the power of poisoning or paralysing marine life, so that infusoria, or vegetable organisms in suspension, coming in contact with the surface, are not held because of its viscidity, arrested by friction, or attracted ; but repelled.

Scum usually originates from the organic or inorganic suspensory matter in water, and which is not in solution, becoming deposited. It may be vegetable slime, have the appearance of slaked lime or pulverised coral, be mud, or eroded matter from the shore or land, cliffs or rocks of the adjacent sea coast, and be assisted, to some extent, by some of the salts in sea water being partly precipitated. Slime may also be dust in a damp state, and frequently consists of a film of microscopic or other plant life. In fresh and brackish water there is usually little fouling, other than slime and grass, although corrosion is severe in brackish waters.

It is known that if it were not for the winds, currents, and tides, even the purity of the sea would not be maintained, as its saltness is insufficient to prevent the growth of fungi and animalcula. Its constant motion prevents this, the spores and germs being tossed about until they are destroyed and eaten by other inhabitants of the ocean, who devour every kind of organic matter which is in the sea. A familiar example of the effect of but few currents in the sea on the formation and growth of seaweed is the Sargasso Sea, the immense tract of weed in the mid North Atlantic Ocean.

It is a characteristic of stagnant water, whether fresh or salt, that it is generally full of life, and, if fresh water, it contains insects, &c., which lay their eggs in the leaves and plants floating on the surface; but as the sea is so vast, that which is appreciable in stagnant fresh water is not so in the sea, because of dispersion, except in certain places. Infusoria, the wonderful and extremely minute enclosed granular gelatinous creatures made known by the aid of the microscope, inhabit stagnant water, whether fresh or salt, in which plants are growing, or in which an abundance of decayed vegetable or animal matter is contained. Their food consists partly of vegetable and animal decomposing matter, and they also prey upon each other. They occur in immense quantities in the water of the ocean, and contribute much to the nourishment of animals of a higher order, particularly in the ocean in high latitudes, where vegetable life ceases to be represented, but animal life is still in abundance. Infusoria are there

Q

found in inconceivable numbers, and form a principal food
of the fishes inhabiting those regions. Some of the infusoria
are able to withstand very considerable changes of temperature without losing their vitality, even from as low as 8° F.
to as high as 260°.

When scum becomes attached to a painted surface it may
dissolve the paint. This may be called the worst case. A
favourable condition may be considered to be that in which
the scum and mud are very thin, and easily washed off,
leaving the surface of the paint but slightly soiled. A thin
scum or layer of mud, if uniform, complete, and not deleteriously affecting the paint underneath it, may act as a protective covering; although any slight usefulness that may
so accrue will probably be more than balanced by the attraction afforded by its formation—for, when suitable soil is
produced for any marine life it will soon discover it and
begin to grow thereon. On a round surface a film or covering
of scum or mud is more quickly deposited, or gathers sooner
than upon pointed or square edges, chiefly because it offers
less opposition to the action of the waves; and therefore
friction, abrasion, agitation of the water, and concussive
action are reduced. Principally from these causes, the forward portions of a ship, except when a vessel remains fast
in harbour, are usually cleaner than those nearer amidships
and aft—especially so in the case of wall-sided vessels. If
a ship's bottom has a semicircular form, which offers the
least friction for a given displacement, it will usually have
more scum or covering of mud on it than one moulded on
triangular lines, which, however, afford the most stability ;
all the surfaces being immersed in the same waters.

Examination by Prof. Tyndall, of nineteen bottles of sea
water, filled under his direction between Gibraltar and Spit-
head, on a voyage of H.M.S. *Urgent*, indicated, as stated by
him at the Royal Institution, that the general tendency of
the examination was to show that the yellowish water of
coasts and harbours held in suspension a large quantity of
particles ; that the particles in the green water were less

abundant and in finer division; and that the blue water of the deep ocean was comparatively free. It is considered probable that the colour of the sea in some places is due to the presence of animalcules or plant life, intermediate between the animal and vegetable kingdoms, for the appearance of the sea varies. It seems white, yellow, of reddish tinge, green, deep bluish green, purple, and even black, the different colours being due to local causes, and the grey and blue tints more especially to sunshine and passing clouds.

The constant roughness or calmness of the sea has a considerable effect upon the magnitude and character of the fouling. Seas possessing powerful and constant currents, especially when carrying sand in suspension and at considerable velocity—as must be the case, or it would sink—have a scouring effect, and, therefore, there will be less fouling than in tranquil waters of a similar nature ; but this may be partly counteracted by the waters of the current being warmer than the surrounding ocean, and by their containing more marine life, as witness the warmer oceanic currents and drifts. Such anti-fouling currents, as they may be called, are, however, generally found in the warmer latitudes, such as those within the Northern Warm, Tropical, and Southern Warm Zones, or between about 35° N. to about 50° S., and sometimes have a large quantity of seaweed floating on their surfaces, which also somewhat neutralises their anti-fouling action as regards a ship's bottom; for these floating seaweed meadows contain countless numbers of marine animals, ready to seek a more stable and benignant surface than that afforded them by floating on seaweed, tossed, and rent about far from the shore. On the Gulf Stream, driftwood has been known to float from the Gulf of Mexico to the western shores of Europe, and may form depositories for crustacea and molluscs, taking them out of the usual range of latitudes in which they are found. A somewhat similar occurrence may be noticed where mounds and hills are formed of the ballast discharged from ships engaged in foreign trade, as at South Shields, for exotic

Q 2

plants may be observed growing on them which are seldom to be seen in this country, except where specially cultivated.

It is important to prevent fresh supplies of water, unexhausted of its combined air, reaching metal, therefore even a film or covering of mud tends to preserve it, and forms a shield to defend the surface from externally active corrosive influences. However, although there is no doubt such a film of mud acts as a preservative coating, it cannot be relied upon for protection, because it is seldom uniform. Careful observation shows that form has a considerable influence in preserving submerged metalwork from fouling and corrosion, for generally on immersed rounded surfaces matter is deposited, but on pointed or square edges such a coating is either prevented, much lessened, or removed by the action of the water, with the result that from the constant renewal and aeration of the water, a certain amount of air in the water becomes absorbed by the metal, which causes oxidation. . Sharp and abrupt edges are also more exposed to the force of the waves or current of water than rounded surfaces, and any protective coating is more likely to be washed off and eroded ; hence the importance of *equal* protection, or the result will be that while the condition of the greater portion of the metallic surface may be excellent, other parts, because of inequality in the protective covering, may be blistered, pitted, have bunches of rust, or be very much corroded ; and as the real strength of a structure is that of its *weakest* part, the mud coating is an unreliable preservative, although, for a time, it may be a perfect shield. This inequality in the thickness of the muddy film, by indicating those places which are least covered, other conditions being similar, will more particularly show where corrosion will probably be the greatest, but the surface will be less foul. No vegetation should be allowed to grow upon any deposit that may have formed upon the metal. The shells and means of attachment of marine animals also protect the surface to some extent, but *unequally*. In order to counteract any such reduction of protection, and the increased abrasion at or about all pointed edges, their additional protection by means of an extra preservative coat, whether paint or composition of any kind, would appear to be advisable.

There is one point to be considered with reference to any protection that may result from the formation of scum on the surface of a submerged structure. It is, will such a film of mud continue to be a preservative, or will it, by attracting marine life of any kind, become so changed or destroyed as to be converted into an active deteriorating influence? It may, under certain conditions; for, when marine plants grow thereon, the scum undergoes an action that may be described as one of combustion, resulting in perforation, however minute, and when that occurs the film loses much of any value it formerly possessed as a shield of protection. The different stages of this scum and mud covering from that of a painted surface being simply covered with white scum only, to the very foul condition of being covered with moss and grass, very thick and long, may be tabulated from the following photographic examples :—

(1) Covered with white scum only.

(2) Moss commenced to grow.

(3) Moss about ¾ inch in length.

(4) Moss and grass about 4 inches in length.

(5) Moss and grass, very thick and long, more than 4 inches in length.

(6) Moss and grass growth on chains after about eight months' immersion.

The mud and scum film is usually from $\frac{1}{16}$th to $\frac{1}{8}$th of an inch in thickness; then moss appears, and may become of any thickness, to about 5 inches or more; and grass and kelp, &c., also grow, and may increase to any usual length. Of course, in different waters, the growth and thickness will vary. Fig. 6 is an example of rapidly-fouling water, but not of any exceptional character.

It will thus be noticed that slimy deposits promote an exuberant growth of moss, grass, &c. It may be stated that, as a rule, when grass or moss has commenced to grow, the anti-fouling paint, or second coat, rapidly deteriorates, and unless the anti-corrosive paint, or first coat, is of very excellent quality, it will be found that corrosion—whether local, in the form of pits, scale, blisters, or bunches of rust, or general over the surface—is in progress; and, without affirming that corrosion will increase as the grass or moss becomes longer, in many cases, either from accelerated deterioration of the paint, in consequence of the plant growth upon it, or from perforation, the plates will be found to be more corroded than if the moss or grass had only just begun to grow. In almost all the instances that have come under the notice of the author, the condition of the plate was good, and apparently free from rust, when the paint was *completely covered with scum only*, and provided the scum had *not* caused appreciable decomposition of the paint, but corrosion seemed to increase *pari passu* with the growth of the grass or moss; and this is in accordance with the fact that all vegetable growth, whether in the state of microscopic fungi, or in that of a visible plant, has a decomposing effect; and, it may be added, the very sources of the food of plants are active agents in the promotion of corrosion and fouling.

If a film of mud or scum did not deleteriously affect the paint, but formed an air and water-tight seal upon its surface, and maintained that condition unalterably, it would be a veritable shield of protection, but it is only a question of time before life in some form or other will be present upon it, and almost from the day of its being a protection, it may be said its powers of preservation decrease, and except, perhaps, at considerable depths, it will become a source from which active deteriorating influences spring. Although undoubtedly a protection under certain conditions, which seldom occur in the waters in which almost all engineering structures are placed, it is so uncertain in character and wanting in uniformity that generally it is unadvisable to rely upon it as a means, additional to the paint, of preservation from corrosion of the surfaces of metallic structures.

It should not be forgotten that although scum in many cases, as has been indicated, is principally marine life in its most minute form, it may also proceed from the slow deterioration, softening, or decomposition, by chemical or other agency, of the paint when in contact with sea water, resulting in the production of a viscous surface, and an inviting resting-place for the attachment, on simple contact, of any suspended matter in the water. Therefore, the surface should be very hard, and not decrease in solidity with age, be even, polished, and incapable of indentation by small floating particles of matter. It should, like Portland cement, have the property of increasing in hardness on immersion, and the surface exposed to water should have no trace of viscidity, but be similar to that of polished marble, although the anti-fouling paint should have great adhesiveness to the first coat, in order that it cannot be recorded of it that it was blistered, thin, and porous, or nearly all dissolved, or much cracked, or loose and badly blistered, or came off on scraping away the scum, or partly or wholly washed off in cleaning, or peeled as soon as the plate was dry, or was nearly all gone, or that there was no indication of any anti-fouling paint.

With regard to the formation of moss, grass, seaweed, and marine plant growth on a submerged structure or a ship's bottom, it may at first be in the form of vegetable slime, which so often covers the surfaces of marshes and pools, and then becomes moss, grass, or seaweed. Seaweed is universally distributed, and best represented in the tropical seas, where the variety is greater, and the masses in which it appears, exceedingly wonderful. It has to be remembered that wherever there is aquatic plant growth, the more highly and delicately formed being found in the warm seas where

" Returning suns and double seasons pass,"

there will most generally be animal life, however minute, in some form or other, attached to it, not a few making it their most cherished home. There are also plants in the sea which have been described as intermediate between the animal and vegetable kingdoms. It is obvious a submerged painted surface must not only have nothing in it to feed or encourage aquatic plant growth, but that it must possess the power to *prevent* plant life attaching itself to it in any waters. This

has been more or less effected by the incorporation of substances known to be fatal to it, and will be subsequently
referred to in Chapters V. to XII.

Fouling by marine plants is caused by microscopic or
minute plants which float upon the open seas, and appear to
colour them, by seaweeds and marine grasses coming in
contact with a ship's bottom, especially if a vessel is engaged
in the coasting trade or in comparatively shallow depths,
because the greenish marine flora found upon rocks or on
ships' bottoms can only flourish in comparatively shallow
waters, for they require sunlight, and it is known that sea-
water does not merely intercept solar light, but modifies its
composition, varying with the depth.   It can, therefore, to
some extent, be ascertained by the colour whether any marine
plants on a ship's bottom grow in shallow water or not.
Seaweeds extend to only a comparatively small depth below
the surface of the sea, generally estimated at from 120 to
300 feet, for they require light.   The variation of the light
is one of the reasons why the grasses and seaweeds having
a greenish hue grow in the shallowest water, and near the
level of the usual highest tides.   In fact, green marine plants
require nearly the same conditions of light as terrestrial
plants ; the olive-green, olive-brown, Indian-red, and red
seaweeds usually being those not receiving so much light,
although there are exceptions.

The spores or undeveloped young plants of Confervæ, and
of seaweeds, move about by the action of their own cilia or
hairs, until they have found a resting-place to which to
attach themselves.   In fact, they have a degree of locomotion,
and, in some cases, may be said to grow, live, feel, and move.
The chief object of an anti-fouling composition is to cause
their non-attachment to a ship's bottom or the surface of a
submerged structure, or, if they attach themselves, to cause
them to die and become detached.   The border line of marine
life has yet to be absolutely defined by the botanist and
zoologist ; however, *all* life must be prevented from attaching
itself.

The life of green plants is dependent upon the provision, in addition to a few less constant materials, of some nitrogenous compound, water, carbon dioxide, and oxygen ; these being utilised only in the presence of sunshine. Any green plant destitute of food stored in its own tissues, if placed in perfect darkness, dies. However, plants can live without sunshine, i.e., by the electric light being substituted for the sun, but they are of sickly, pitiable appearance, and, although they turn to the light, they have but a fragile existence. Covered or subterranean reservoirs for waterworks have been in use from the earliest times, and it is probable they were so made not only to keep the water cool and free from the impurities of the outside air and dust, but also to prevent the growth of algæ, which may generally be almost entirely prevented in most water by a total absence of sunshine. Reference is here made to this because a remembrance of the degree of sunshine experienced on the coast, in any but the most arid and barren districts, and the warmth of the sea, will enable a very good idea to be formed of the degree of fouling and corrosion that is likely to be experienced, for plants which in ordinary circumstances grow very slowly, develop with extraordinary rapidity under the influence of a warm and humid atmosphere. Seaweeds being carried by marine currents are frequently encountered at considerable distances from shore, and are found to be inhabited by myriads of little crustaceans, which may be so transported to great distances from their birthplaces. Most subaquatic life is very persistent; for instance, in addition to the growth of the spongilla in lakes, reservoirs, &c., the *Spongilla lacustris* occasionally grows in water pipes by pieces finding their way into the pipes from the source of supply. The sponge is an animal, and lays eggs, which attach themselves in large quantities to the interior surfaces of the pipes. Flushing will not remove them, scraping or wire-brushing being necessary. Treatment of the water obtained from subsoil some seven miles away, originally used for the water supply at Konigsberg, which was very rich in iron com pounds, so as to be almost useless for domestic services, appeared to show that the existence of iron salts to the extent of four parts per 100,000 in it contributed greatly to the growth of algæ, which at length filled the mains. First, a species of spirillum appeared, succeeded ' by the leptothrix, subsequently replaced by the crenothrix. Allowing the water to rest for a time did not succeed, although deposits of mud where removed when the subsiding beds were cleansed. The remedy adopted, after careful investigation, was to convert the soluble iron salts into insoluble compounds by means of careful aeration, withdrawing thereby also the food stuffs of the various algæ. This was done by simply turning the water in an ancient watercourse for a length of five miles. It was then passed into filter beds, and made fit for domestic supply. Bacteria are the most numerous on the surface. They are present where putrefying matter is found, as in river and lake water, and, by attaching themselves to sandy or loamy particles, form a continuous covering, as do the algæ and the diatoms, and the process of deposition of slime continues by these and various micro-organisms until a slimy coating is formed. Such a covering, if it does not harden, will yield to a current of water having a greater force than that at the time of the deposition of the slimy life-matter. Experiments conducted under the guidance of the Hygienic Institution of Berlin University show that the effect of sand-filtration on river and lake waters, in which micro-organisms abound, is due in but a slight degree to chemical action, and almost exclusively to the labours of the bacteria; and that when sand is subject to water passing through at a rate of 4 inches vertically per hour for two hours, the sand, in itself almost neutral, is merely a matrix in which the bacteria bed themselves, multiply, and perish, thus causing the slimy coating of the sand grains. It is well to remember this in determining the probable causes of fouling. In canal beds—whether the water is salt, brackish, or fresh—in warm climates, plants of a weedy nature grow quickly and luxuriantly in a tangled mass.   The prevention of all subaquatic plant growth on the surface of a structure is of importance, for marine animals are attracted by it; it is an obstruction, and tends to deteriorate the paint, for all vegetable life upon it, whether in the state of microscopic fungi or in that of a visible plant, has a decompósing effect.   The chief aim is not to permit the spores of such life to find a congenial soil, or one upon which they can germinate, and also to prevent the particular conditions under which they each exist. Poisons peculiarly destructive to vegetable life have, therefore, been incorporated with the pigments or substances forming paint, for the formative fluid must be destroyed in any vegetable matter that may become attached to its surface, or growth will result.   In fact, the paint must inseparably contain or be incorporated with some substance upon which no marine life, whether animal or vegetable, can exist or become attached.

237

CHAPTER II.

237

FOULING BY MOLLUSCA, CORAL, AND MARINE ANIMALS.

WITH regard to fouling by mollusca, coral, and marine animal life, the scum or mud on the surface of paint on a submerged structure is often full of infusoria, especially when it has the appearance of slaked lime or pulverised coral, for they live in infusions of animal or vegetable matter. It is necessary to briefly consider their nature, and that of a few of the animals which adhere to a sub-merged surface, in order to prevent their attachment. In-fusoria do not possess a nervous system. Their most usual mode of motion is by fine lippets or hairs, which exist either scattered or arranged in regular series over the whole surface of the body, or are collected in considerable numbers round the orifice of the mouth. They are movable at the will of the creature, and serve either as organs of locomotion, or for the production of whirlpools or eddies in the water, by means of which the minute particles the animal eats are brought within reach. In some these ciliary hairs become converted into movable bristles and hooks, which enable the animal to crawl upon fixed objects in the water, and even to execute distinct leaps. By their incalculable number they amply compensate for their want of size; some are said not to exceed the $\frac{1}{20000}$ of an inch, the largest being about $\frac{1}{30}$ of an inch in length. However, almost every drop of water on the face of the globe appears to contain them in greater or less profusion, and this, coupled with their great fecundity—it having been calculated that the progeny of some animalcules would amount to 268 millions in four weeks—demonstrates the vast quantity of food furnished by these creatures to others a little higher in the scale, which in turn become the prey of larger animals.

The sense of smell in all the lower animals is a very important one. It has been well called an anticipatory taste. They judge by it of the fitness or unfitness for food of any object they happen to find. Carnivorous fishes, like the sharks and rays, it is known, have a highly developed sense of smell, and sniff from afar the presence of anything fitted for the food of their own species. They have followed a ship for miles attracted by the odour of meat or of human beings. Whether the minute marine animals which partly cause fouling on ships and structures have also a delicate sense of scent or smell, authorities on the subject must finally decree, but, from experiments conducted in the presence of the author of this book with an exceptionally powerful anti-corrosive and anti-fouling substance, possessing singular merit and great power, undoubtedly the little marine animals could be seen to closely approach the coated surface, without apparently touching it, and then to quickly recede. Some, who probably were of a bolder nature, touched the surface and seemed to become paralysed, or subject to a shock which effectually prevented their attachment to, or fouling, the coated plate.

Mollusca may be defined as animals possessing a body which is soft and fleshy, devoid of bones or any internal skeleton, and not divided, as in insects and worms, into rings and articulations. This body is covered with a soft, irritable, and muscular skin, and is moistened by a viscid liquor which exudes from it. It is in very many instances of sufficient extent to form fleshy folds that envelop, more or less completely, the whole body, as in a mantle or cloak. In some cases this skin is naked, and then the mantle is thick and mucous; in the greater number of the species, however, it is protected by a hard, persistent, calcareous covering, called a shell, and then the mantle is thin and transparent. The word mollusca is now adopted to contain

all animals which are true shell-fish, whether they have or
have not shells, and many have no shells. The most
essential character lies in the arrangement of their nervous
system. This consists of a certain number of enlargements
or centres, called ganglia, from which the nerves are given
off to the different parts of the body. These ganglia are
principally concentrated around the entrance to the alimen-
tary canal, and form a collar or ring which surrounds the
throat, and are connected with other ganglia disposed
without symmetry among the viscera, or in the neighbourhood of the organs of locomotion. In referring to the
general characteristics of mollusca, so far as affects their
attachment to submerged structures and ships' bottoms, it
is necessary to cursorily anatomise them. The molluscous
division of animals consists of creatures whose bodies are
universally of a soft consistence. Most of the mollusca
possess special organs of touch, in the shape of tentacles,
arms, or lobes, situated on the head or in the neighbourhood
of the mouth, or of cirri upon other parts of the body. The
skin, which is always soft, apparently has great sensibility.
They also seem to possess auditory organs, and the senses of
smell and taste appear to be exercised to a certain extent by
most of these animals, but whether any organs specially
devoted to these functions really exist is doubtful. The foot
consists of a mass of muscular fibres, by the contraction of
which, movement is affected. In a great number of mollusca
the foot forms a flat disc, which adheres to any substance
to which it may be applied. In others, the foot is bent
upon itself. In the burrowing species the foot is the
agent by which they bury themselves in the sand or mud,
and in those that bore into solid rock, the surface is covered
with minute silicious particles which they deposit. Others
are confined to a single spot, as the oyster, except during the
earliest period of their existence. The foot in them is
either wholly undeveloped, or serves to support a glandular
organ, from which a silky matter is secreted, which serves
to attach the animal to submarine objects, as in the mussel.

240 PREVENTION OF FOULING AND CORROSION.

In the highest class of mollusca, the Cephalopoda, the mouth
is surrounded by a variable number of arms, covered on their
inner surface with numerous cup-like sucking organs—as
in the *Octopus Hawaiensis*—which serve as organs of motion
and for the capture of prey, and enable the animal to take a
firm grasp of any object. They are enclosed within a soft
flexible skin, called the mantle, possessing great contractile
power, and most of them are protected by a hard calcareous
covering or shell, which is secreted by the mantle, and is
gradually increased in size, in proportion to the growth of
the animal. The bivalves, or two-shelled with a hinge,
exhibit no traces of any distinct head, but in the univalves,
or one-shelled, this part of the body is well marked, and
usually furnished with special organs of sense, tentacula,
eyes, &c. The shell consists principally of carbonate of lime,
with a small quantity of animal matter. The calcareous
matter is deposited in the cells of the edge of the mantle,
which are in contact with the free margin of the shell. The
shell is almost always coated with a layer of animal matter,
the epidermis, of greater or less thickness. It is of a horny
consistence, and serves to protect the shell from the action
of the carbonic acid, which is often dissolved in considerable
quantity, especially in fresh water. In many places, however, the epidermis is an insufficient protection against the
corroding action of the water, for it often eats deeply into
the substance of the shells of the mollusca which make it
their habitation. Carbonate of lime, in some, helps to
support their gelatinous frame, as in the third order of
Zoophytes, i.e., animal plants, to which belong the corals.
They form for themselves a permanent attachment to a
substance by a deposit of calcareous matter. The shells of
Crustacea and the Cirripeda are chiefly made of carbonate of
lime held together by animal matter.

The class of Bryozoa grow on a common stock; the cells
are sometimes horny and sometimes calcareous. They frequently stand upon short footstalks, rising from a tubular
stock, which creeps over the surface of stones and aquatic

plants. In other cases, the cells are sessile, forming a crust upon submarine objects, whilst in others the colony is attached only by its base, with the opposite extremity floating freely in water. One division of them have the tentacles placed in a circle round the mouth ; these are all marine; the others, which inhabit fresh water, have them upon two arms given off from the sides of the body. Many grow upon the fronds of seaweeds.

The shells of mollusca often consist of about 96 per cent. of carbonate of lime, 1½ per cent. of phosphate of lime, 1½ per cent. of water, and only 1 per cent. of animal matter ; and in some shells, as oyster-shells, scarcely an appreciable quantity of animal matter. The stony cases of the Polyps, or corals, are also mineral bodies, as they contain from 97 to 98 per cent. of carbonate of lime, besides small quantities of magnesia, alumina, iron, and silica, combined with phosphoric and fluoric acids. Shells of mollusca are found in all stages of the petrifying process, i.e., from slight discoloration to a state of entire transformation into crystal carbonate of lime. Petrification may proceed from several causes, which have thus been well summed up by Major-General Portlock, R.E.: by abstraction of matter, by incrustation, by the mechanical introduction of sedimentary matter, by molecular penetration, and by substitution by chemical change, and by transformation. Although on a ship's bottom petrification, owing to want of time, can seldom occur, in fixed structures in the sea it can ; and it is well to give it a thought in considering corrosion and fouling.

Some marine mollusca are littoral or shore inhabitants, others exist almost entirely at various depths below the sea-level, and those which are found in calcareous rocks excavate chemically by a dissolving acid solution ; hence, fouling by mollusca will vary in extent and character. Most oceanic and marine organic life can alone thrive under certain conditions suited to its individual existence, so that all such animals are not found in the same place, or seas. In some of the warm seas countless myriads of minute little insects are

R

producing or building rocks which defy the power of the waves, and are slowly rising up from the bottom of the sea, as in the coral lands of the Pacific. Branch coral is often found attached to a ship's bottom when a vessel trades in warm waters.

Animals of the Tunicata class generally present the appearance of shapeless gelatinous masses. They are all marine, and are numerous in all seas, being found from low water to a considerable depth adhering to rocks and sea-weeds. In the northern waters they are sombre in colour, but in the sunny regions of the south their hues are very bright. Some are minute gelatinous creatures, such as the Salpæ, found in all seas, but most plentifully in those of tropical climates, which are often filled with them in such numbers that a ship passes for days through masses of them, and at night they are exceedingly luminous. The Ascidiæ, order 1 of the Tunicata, are either attached by the base to submarine objects, or united together in various ways; some by means of a creeping stem, that runs along the surface of submarine objects, and gives rise at intervals to short foot-stalks, at the extremity of which the animals are supported. Others live separately, and are attached by their base to submarine bodies. Some are in the form of a cartilaginous tube, open at one end, as the Pyrosomatidæ. These latter animals are found in the seas of warm climates, where they float along in an upright position. Like all the Tunicata they are luminous in the dark. Their food consists of minute particles of organic matter contained in the water.

It is said the cause of the phosphorescence seen in the English Channel on certain occasions is the presence of multitudes of the minute *Noctiluca miliaris*. The *Nereis noctiluca* is also an inhabitant of every sea, and is an active agent in the production of phosphorescence. The luminous principle has been declared to be a fluid organic substance, called Noctilucine by Dr. Phipson, who has specially investigated the mattter. It is produced by both aquatic and land animals, and by mollusca, vermes, crustaceæ, and other curious zoophytes.

The Pteropods are all of small size, and are inhabitants of the ocean, and seldom venture near the shore. On the high sea they often abound in such profusion as to colour the surface for miles. The species of the genus Clio, belonging to the order Gymnosomata of the Pteropods, are found in marvellous numbers, principally in the Arctic and Antarctic seas, and form one of the principal sources of the nourishment of whales.

The edible oyster attaches itself by its left or convex valve to rocks or other submarine objects. The fry consists of whitish gelatinous masses; these soon fix themselves by the shell to some substance. The Cockscomb oyster, inhabiting the Indian Ocean, however, attaches itself by means of a kind of plug, which passes through an aperture in one of the valves. Oysters, particularly in the warmer waters, will attach themselves to a ship's bottom or submerged structure; and mussels also, but by means of a byssus. They are generally marine, although the common mussel may be found in fresh water. They inhabit most seas.

So numerous are the different mollusca that but a few can be here referred to. From the brief descriptions given, it will be noticed they attach themselves to objects in different ways. In the Eastern Ocean, and in the seas of the warm parts of the world, they are found most abundantly, the Indian and China seas being especially prolific, and also the waters of West Africa, the West Indies, and the Persian Gulf. The habits of several are not thoroughly known. As mollusca feed upon minute marine animals, it is necessary to prevent any life, minute or not, attaching itself to the surface of a submerged structure, or a ship, in order to keep it clean. Of all marine animals, so far as the fouling of ships' bottoms is concerned, the Cirrhopoda are perhaps the worst, the common barnacle being the best known. It generally attaches itself to floating objects, and frequently covers the bottoms of ships to such an extent as to impede their progress through the water. Cirripeda are now universally admitted to be crustaceans, though they used to be classed with the mollusca, because of their shells. They are articulated animals, contained within a hard covering composed of several pieces, and consisting of laminæ of calcified chitine. They are divided into two families: the Lepadidæ, or Barnacles, and the Balanidæ, or Sea Acorns. The first, the barnacles, are characterised by having a lengthened peduncle, which is coriaceous, hollow, flexible, and provided with muscles, by means of which they attach themselves to submarine bodies. The species are numerous, and are found extending over the whole world, the greater number, however, inhabit the warmer, temperate, and tropical seas. In their young and free-swimming state they float freely in the sea, swimming rapidly with their backs downwards. They soon become invested with a coriaceous shell, and they now begin to look out for a fit place to which they may attach themselves. Having found this, they adhere to it by the part of the body near the head, and in a few days afterwards they throw off their external covering, and then their growth is very rapid. The shelly valves form, the antennæ disappear, being cemented to the surface of attachment; the eyes remain, but in a more or less rudimentary form ; three more pairs of legs are developed, and the animal assumes its perfect state. The barnacle may then be said to be fixed for life, head downwards, and it receives a mouth and stomach, which it had not when in the free-swimming state. But all barnacles do not so change, for the males of some kinds live inside the conical shells which hold the female, and do not have their own shell, for they remain in the free-swimming state. When a barnacle is developed, curled hairs, to the number of twelve pairs, are constantly being protruded and withdrawn through an aperture in the shell. They are the feet of the animal transformed into these organs, and by means of them a current is produced in the water, which attracts to the mouth the objects which serve as food. Their nervous system is ganglionic; they have a double circulation of blood, and respire by means of branchiæ or gills, the form of which varies. The Balanidæ, or Sea Acorns, are characterised by their being fixed to submarine bodies without the intervention of a peduncle, which the barnacle possesses. The body of the animal is enclosed in a bell-shaped tube, open at the top, and formed of several valves united together, side by side, by a dentated suture, and covered by an operculum, which consists generally of four valves nearly equal-sized, and sunk into a flexible skin, allowing them a considerable degree of motion. The base by which they are fixed is a shelly plate. The species are numerous and most widely diffused. They attach themselves to ships, rocks, timber floating or at rest, shells of crustacea and mollusca, &c. Some are of considerable size, and one species of the genus Balanus, found on the coast of Chili, growing to a size of 5 or 6 inches, forms a very common and highly-esteemed food for the inhabitants. Another is equally prized by the Chinese. Some of the species are found attached to sponges, while others are embedded in corals. A [shell of a balanus, or sea acorn, taken by the author from one of the bottom plates of a ship, weighed 4 ounces, and its attachment covered a space of 2 inches square. The length or projection of the shell was 3 inches; and the thickness from $\frac{1}{32}$ to $\frac{1}{8}$ at the end to $\frac{1}{4}$ to $\frac{3}{8}$ of an inch at the point of attachment. If 1 square foot was closely covered with them, it would contain $\frac{144}{4} = 36$, but an allowance must be made for the peculiar shape and irregularity in form of the base of about 33 per cent.; therefore the number would be 24 per square foot; and as, in this instance, an empty shell weighed 4 ounces, the weight in lbs. would be $\frac{24 \times 4}{16} = 6$ lbs. per square foot, *without* that of the animal inside it. Their weight, which, if a ship's bottom were encrusted with them, would be from 30 to about 100 tons, although decidedly detrimental, is not so much the cause of impediment to a vessel's progress, as the obstruction offered by the roughness and irregularity of the surface, and the consequent increased friction.

In temperate climates, the season of the year has an effect upon the degree of fouling, for coast marine animal life goes down to deep water in winter, and seems to disappear or sleep. In the winter months, the shores, or the sea contiguous to them, no longer afford food, so the animals go to deep water, where life is not subject to such climatic conditions.

The drift wood which the Gulf Stream floats from the Gulf of Mexico even to the western shores of Europe, is found to be frequently perforated by the larvæ of insects, and may serve to carry the eggs of fishes, crustacea and mollusca, and so cause unexpected fouling in certain localities.

CHAPTER III.

FOULING, AND THE TEMPERATURE AND SALTNESS OF THE SEA.

THE temperature and saltness of the sea, especially the former, have to be considered in estimating the probable magnitude of the fouling, and, to some extent, the corrosion. In north glacial, or Arctic latitudes, marine worms and minute crustaceans are found. In the northern temperate, say from about 55° N. to 35° N., all classes of mollusca are present, and numerous polypi, though the corals proper do not yet appear in abundance. In the northern warm, from about 35° N. to 30° N., in a less degree, and in the large region called the tropical, from about 30° N. to 25° S., the marine animals, as a whole, are as equally superior to those of other regions as most of the other types of animals. The polypi, for instance, here display an activity of which the other zones present no example, and whole groups of islands are covered with coral reefs formed by these little animals, hence the name, " The Coral lands of the Pacific." Thus the effects of aqueous temperature on the inhabitants of the ocean are very marked, not only by the growth of coral in the tepid currents, but also by the presence of thousands of shelled mollusca of gorgeous colour and enormous size. Especially in parts of the Indian Ocean and China seas are mollusca numerous and remarkable. In fact, each division has marine animals wholly or for the most part peculiar to it, and, even when placed under the same circumstances, those of the eastern and western hemispheres are not precisely alike within the same latitudes. Briefly, the ocean may be said to be full of streams, each with its own course, temperature, aquatic fauna and flora, and therefore peopled with its own denizens.

The different degrees of saltness and temperature of the sea also bear testimony to the variableness of the conditions to which a submerged or floating structure is subject. For instance, the waters of the port of Cronstadt are almost fresh, caused principally from their being nearly land-locked. The Baltic Sea has but about one-fifth to one-sixth of the saltness of the English Channel, and the waters of the Cattegat and the Sound about one-half. On the other hand, the North and South Atlantic, Sea of Marmora, and the Mediterranean may be considered as having from 11 to 20 per cent. more saltness; the Black Sea being the exception among the southern waters, owing principally to its land-locked or interior position, for it possesses only about one-half of the salineness of the English Channel waters; but the Red Sea, which is exceptionally subject to evaporation, and is contiguous to a riverless district, contains nearly 30 per cent. more. The amount of saline matter in the open sea is greatest at a distance from the shores, for there the salinity of the ocean waters is not diluted by the volumes of fresh water discharged into them by rivers. As a general rule, the greater the evaporation the more salts in the sea water : this is shown by the difference between those present in the torrid zone and Arctic and Antarctic seas. In 1000 parts, the salts present in sea water may be taken, according to some analyses, as about 0·61 of a part at Cronstadt; 5 to 6 parts in the Baltic Sea; 16 parts at the Cattegat and the Sound; in the Black Sea, from 16 to 21 parts; German Ocean, 32 to 33; in the Indian Ocean, between Africa and the East Indies, and the great Ocean between the East Indies and the Aleutian chain of islands, 34; North and South Atlantic, between Greenland and Cape Horn, about 36 to 38 ; Mediterranean, about 38 to 39.

The composition of sea water, obviously, must vary considerably in different places, and is also influenced by the weather, whether calm or rough, fine or wet. In 1000 grains of sea water of the English Channel, about 28 grains are common salt, and about 8 grains other substances. An analysis of such water gave: water, 963·74; chloride of sodium, 28·06; chloride of magnesium, 3·67; chloride of potassium, 0·77; sulphate of magnesia, 2·29; sulphate of lime, 1·41; carbonate of lime, 0·03; bromide, 0·03; with traces of iodine and ammonia. In the Baltic waters, about 84 per cent. of the salts contained in the water have been found to be chloride of sodium, and about 9·5 chloride of magnesium; whereas in the North Sea, the chloride of sodium has only been about 75 per cent. of the salts.

Chloride of sodium, or common salt, is the source of muriatic or hydrochloric acid, which is generally present in a marked proportion in the air of a sea-shore. Iodine exists in sea water, its chief source being the oceanic algæ or sea-weeds. Bromine has also been found in marine plants, and in sea-water, combined with the base of magnesia. In the ashes of seaweed and plants growing on the sea-shore within reach of sea water, carbonate of soda exists. It has been considered probable that the saline condition of the sea is the result of long-continued action of water upon the solid mass of earth. The sea is constantly more or less changing the form and position of a shore by either eroding or depositing matter. In shore or littoral deposits, also in shallow or deep-sea and other marine deposits, the organic remains and the rate of fouling and corrosion will vary. As the waters usually contain from about 3 to 4 per cent. of various mineral salts, which have been previously described in the analysis, the formations produced are often more fixed and solid than those of fresh water. This is especially the case in warm climates. The approximate values of comparative saltness of different seas are here mentioned, because, although they do not greatly influence the magnitude of the fouling, they have an effect as regards the corrosion of submerged work.

It may here be well to bring to remembrance that the reason sea water is brackish is because it holds in solution a quantity of saline matter of which common salt is the

principal ingredient; and that, although rain water is largely
evaporated from the sea, it is not salt, because the saline
matter does not evaporate, consequently when sea water is
converted into vapour, and forms clouds, or is in the air, the
salt remains in the sea.

The temperature of the water has a greater effect on
marine life than the saltness of the sea.   It is much more
uniform than that of the atmosphere, water being a slow
conductor of heat, and not being subject to rapid changes
like the atmosphere.   Its all-ruling influence upon every
living thing is well known, and is shown in aquatic growth
and the inhabitants of the sea.   Uniformity in fouling may
extend along a coast similarly and equally exposed to the
various natural agencies, but so soon as differences of climate,
temperature, water and currents occur, it will not be of the
same character or quantity.   Here the temperature at great
depths need not be considered, for ships and engineering
structures are placed in the surface waters, with but few
exceptions, such as submarine cables, anchors, and chains.
The temperature of surface waters varies considerably, but
not absolutely according to the latitude, for in some latitudes
it is modified by ice drifted by the polar currents from the
Arctic and Antarctic Oceans, and also by the Gulf Stream,
and the other ocean currents, especially those in the torrid
and north temperate zones.   It varies very considerably, and,
as regards oceanic waters, from about 86° F.—the greatest
heat of the oceanic water of the Mexican Gulf, or Gulf
Stream, which is about 9° above the ocean temperature due
to latitude alone—to as low as about 31° F., which has been
registered in latitude 60° N. in March, omitting consideration of the Arctic and Antarctic Oceans as being, at present,
without the circle of active constructive engineering operations.   At about latitude 60° N. the temperature of the surface water during summer, no ice being near, has been found
to be about 50° F.; at about 50° S., 37° F.   The depth of
water and nature of the bottom of the sea affects the temperature; for instance, water over a sand-bank is colder than

where it is deeper. Although the temperature of the earth increases as the depth below the surface becomes greater, the temperature of the sea *decreases*, and this is so at the tropics.

As temperature determines in great measure whether mollusca, plant, and all marine growth will be present, it is always advisable to know the average temperature of the water, as it will indicate the kind of mollusca and other marine life which may cause fouling and corrosion. In the warm seas, coral and thousands of shelled mollusca flourish, which would shun colder water as much as whales do warm water. The inhabitants of the Mediterranean Sea, which has a temperature usually four or five degrees above that of the external ocean, are, for the most part, of different species to those of the ocean. In the China seas and Indian Ocean, and also in the Yellow Sea, away from the colder currents encountered near Japan, although only a few hundred miles apart, mollusca will live that either cannot or will not live in the colder waters. The sea near Java, Malabar, Madras, Kurrachee, Ceylon, Mauritius, the Philippine Islands, the African coast, particularly the western and southern parts, and the Brazilian and West Indian ports, and those in the Gulf of Mexico, are generally regarded as very quickly-fouling waters.

CHAPTER IV.

NOTES ON CORROSION IN SEA WATER, AND THE INJURIOUS EFFECT OF FOULING.

THE general causes and action of corrosion are examined in the First Part of this volume. It is agreed that the corrosion of wrought iron, steel, and metals in sea water arises from the chemical influences of the saline constituents of the water, aided by the action of the contained carbonic-acid gas, oxygen, and other gases, and is increased by galvanic action, which may arise in various ways. Simple corrosion may be said to occur when a piece of iron is free from any special interfering or accelerating conditions, but inequalities in the mass may lead to galvanic action, as also particles of magnetic-oxide scale on the metal.

What might be termed compound corrosion is due to the joint action of simple corrosion—as described in the First Part of this volume, on ' Corrosion and its Prevention, and the Safety and Preservation of Iron and Steel Structures '— and to that of galvanic action.

Corrosion on the plates of ships or submerged iron or steel structures seldom occurs *equally* over the surface, but is usually in very small rust spots, which may increase to about $\frac{3}{8}$ or $\frac{3}{4}$ of an inch across; in pits, at first superficial, but gradually becoming deeper, until at length perforation takes place ; or in blisters or bunches, which may cover the surface like fine bubbles, or be of any size from $\frac{1}{2}$ to 1 inch across, or even more. It is usually more on the inside of the plating on some parts of a vessel's hull than on the outside surface

of the plates, and an adhesive, hard, and effective anti-
corrosive paint is a necessity, in order to preserve the plates
from rust; but this may occur from want of continued attention, as the internal corrosive influences can hardly be
compared with those on the outside surface of the plating,
except, perhaps, where bilge water has free action on the
rivets and plates, and in and about the bunkers, in some
measure consequent upon galvanic action set up by the coal
dust, in and about the engine and boiler spaces, and peaks,
and in the plates of the hull which are near the boilers.

The cellular double-bottom system of building ships has
decided advantages in accessibility for examination, cleaning,
and painting ; and therefore corrosion in the inside of ships
so built can be more readily prevented or arrested. Saline
particles are not evaporated in the holds of ships, nor are
they exposed to the drying action of currents of air; and the
salts in sea water being very hygroscopic, i.e., readily absorbing and retaining moisture, have an energetic corrosive action
on metal. Draughts of air, when spaces must be nearly
closed, generally help to lessen the rate of corrosion, because
of their drying effect, but not always, for the air may be
almost saturated with moisture. In a ship or pontoon,
dampness and moist warm air must usually be present, and
therefore exposed ironwork, such as iron bolts, corrode; and
rust very deleteriously affects wood in contact with it.* As
iron contains sulphur, it has been stated that the sulphur
becomes converted into sulphuric acid, which decomposes the
salts present in the sea water, and in oak, &c. Neither the
salts nor oxide of iron prevent fungus growth. Galvanising
does not then protect iron bolts from corrosion, as the acid of
oak and other woods gradually dissolves the zinc coating,
corrosion follows, and also decay of the wood.

With regard to the points of greatest corrosion, the
different effects of still and moving water upon the rate of
corrosion may be best shown by examples; namely, that

---

* See Part I. of this volume, Chapter XVI.

corrosion, except under peculiar circumstances, usually increases externally in ships upwards from a few feet above the keel, the worst places being near the wind-and-water line, and a few feet down, i.e., where the plates are alternately wet, dry, and exposed to the air. Experiments and experience have proved that waters constantly aerated or changed by currents, or disturbed, such as the surface waters of seas and rivers, continually supply fresh quantities of active corrosive agents; both iron and steel consequently become corroded much quicker than in still water. The constant wash of free bilge-water in ships, when the internal plates are not properly protected, also, soon corrodes and destroys the heads of rivets, and grooves the plates between the frames on the path the water takes. In iron-pile structures in ordinary sea or brackish water, and in a lesser degree in tidal river-water, such as the Thames, examination has shown that corrosion appears to be greater in cast-iron piles at or about low water, and apparently does not energetically extend much below that level, or above where the water is well aerated or the pile alternately wet and dry, but the effects of spray may somewhat modify this, as also places where the water is impregnated with sewage, which may be particularly present near the ground. It was considered in some experiments that they showed the corrosion of Yorkshire wrought-iron by sea water in the open sea to be seven times greater than in still sea-water.

Examination of the iron plates of an 'ocean liner, after about fifteen years' service, appeared to show that the corrosion increased from about the 4-feet water line, to which point it was inappreciable, nearly in the proportion of the square root of the draft to 16 feet up ; that is, at the 6-feet waterline, it was about 20 per cent. more than at 4 feet; at 12 feet up, 70 per cent. greater ; at 16 to 20 feet, about 100 per cent.; but it did not seem to increase above that line, and between the 6 and 12-feet lines, in many places, the increase was not more than that between the 4 and 6-feet lines, but *above* the 12-feet, corrosion increased quickly, and was equal at

16 feet to 40 per cent. of the original thickness of the plate above the turn of the bilge. However, this is only referred to as a single instance, and more to prompt those who may happen to read this to record their experience, as, by comparison of the condition of a large number of hulls, it is possible a reliable indication of the comparative corrosion under similar conditions might in time be obtained, although the preceding results are believed to be not unusual.

In order that the speed of a vessel may not be lessened, weeds, barnacles, and all marine life and growth must, as far as possible, be prevented from attaching themselves to a ship's bottom. In some warm waters, a thick coating of grass may adhere while a ship is being loaded, unless the plates are coated with a very powerful anti-fouling composition. When a steamship has great speed, the rapid motion will almost keep the surface free from weeds in many seas, especially when steaming through any swift currents in which sand is carried; however, although sand currents may somewhat clean a ship's bottom, external friction may soon abrade any paint on the plates, and cause thin or weak places at which rust spots will be formed, and then friction or pressure will accelerate the corrosion. To maintain a clean bottom on a ship, or one free from much fouling or any but slight scum, is a great advance towards the realisation of the success of a ship either financially or scientifically. Numerous instances have occurred in which it was not possible to attribute the reduction of a ship's speed by several knots except to fouling. That the speed is considerably reduced is generally acknowledged; however, it is not easy to reliably compare the steaming of a ship with the bottom foul and clean, under precisely similar circumstances of weather, load, and steaming, for they so seldom occur; but the diminution in speed, caused by an ordinary foul state of a steamer's bottom appears to range, and some will say it is much more, from about 1 knot upwards. In ships of the mercantile marine, consequent upon the ever-varying conditions of cargo, draft, steaming, weather, build of ship, degree of fouling, and a variety of other causes, a reliable comparison is difficult to attain. However, fortunately, such interfering conditions scarcely exist in the navy, and the very highest authority, Sir Wm. White, stated in his recent paper before the Institution of Naval Architects, that " Careful observations made in the Royal Navy, on ships employed in European waters, have shown that, after five to six months afloat, about 20 to 25 per cent. more power was required to maintain ordinary cruising speeds than sufficed with clean bottoms; after ten to twelve months afloat this increase of power became 40 to 50 per cent. The ships on which these figures are based were not considered exceptionally foul when docked." It is certain the maximum speed is considerably reduced, and that the retarding effect is much more than is generally thought, and may be anything from 5 per cent. to even 50 per cent. It is said Edison has found in experiments that the friction of salt water and its constituents on the sides of ships is much more than is generally believed. The importance of a thoroughly preservative anti-corrosive and anti-fouling composition, possessing a smooth and durable surface, can hardly be exaggerated, even when only regarded from the point of view of loss of speed.

CHAPTER V.

PREPARATION OF THE SURFACE OF METAL BEFORE PAINTING.

BEFORE proceeding to consider the preservation of a metallic surface by paint or composition, it will be well to refer to the preparation of the surface of iron or steel so that it may be in a fit condition to receive the preservative coat. It is very important that the surface should be in such a state that only the pure bare wrought iron or steel receives the preservative coat, but not in the case of cast iron, in which the silicious skin is a protection; for if there be any other substance upon the said metals, it is covered by the paint, and not the metal, and the paint may therefore be spread upon a corrosive substance which, sooner or later, will cause loosening of the coat of paint, and ultimately an opening for the entrance of corrosive influences, which will then probably gradually attack the whole surface, and render the paint useless. Cases occur in which paint is spread upon so dirty or unkempt a surface that the best composition is so wrongly used that it is not given even a chance to prevent corrosion and protect the metal. Again, it may be spread upon a decaying coat of paint under which corrosion of the metal has commenced, whereas the previous coat should be removed. These simple precautions, although they obviously cause a little increase of labour, soon repay their cost with interest, in the freedom of the metal from corrosion and the prolonged serviceable life of the paint.

The proper preservation of the scale or skin on cast iron, and its removal from wrought iron and steel, has been

S

examined in Chapter **IX.** of the First Part of this volume. That may be said to be the first operation. The second is to completely clean the surface. First, the preparation of surfaces that may be called unsubmerged will be considered, although all exposed structures are practically in a state of submersion when rain falls; and, secondly, submerged structures, which are subject to fouling as well as to corrosion. To properly prepare the surface of wrought iron and steel, especially steel, before painting, the chief essentials are: (1) to remove the scale they receive during manufacture; (2) to neutralise any acid or other solution used in effecting it; (3) to make the metal perfectly dry and free from corrosion; (4) to cause it to be clean, even, and without dust or any other impurities. In the case of cast iron, the scale being entirely preserved, requirements Nos. 1 and 2 do not apply, but Nos. 3 and 4 do.

All the scale on wrought iron or steel, and which is present on the manufactured metal, should be removed from the surface before painting operations are commenced, or it will fall off with the paint and leave the sound metal bare and exposed to the atmosphere. Immediately the scale is properly removed, and the surface of the plate dry and efficiently cleaned, ready for painting, a coat of anti-corrosive paint should be applied. The iron used for bridge construction on the Bavarian State Railways was treated in the following manner before being put together: it was cleaned, then dipped in dilute acid, washed in lime water, well rubbed until clean, immersed in boiling water; and when it had acquired the temperature of the water, it was taken out, painted with hot linseed oil, and then with non-corrosive paint. All the ironwork of the Isar Bridge, near Munich, before being erected, was cleaned in dilute acid and scrubbed with iron scale, after which it was steeped for from five to fifteen minutes in boiling linseed oil, and then painted with red lead. Pickling in acid, neutralising, washing and drying, is the best method of preparing the surface of iron or steel for painting yet practised. Answers from thirty-nine railway companies to questions put by the German Railway Union concerning painting, were, that in new work, the iron is first cleaned, in some cases by scraping or rubbing, in others by plunging first into acid, and then into lime water, and finally it is painted. The iron and steel work of the Kuilemburg Viaduct, Holland, were cleaned in a bath of muriatic acid, and properly washed and dried before painting.

An old practice is to require the plates, after leaving the rolls, and at a certain temperature, to be immersed in oil. This is of doubtful utility, for the oily surface attracts dust and dirt, which is not easily removed, during manipulation, fitting, and riveting; and workmen naturally dislike to handle such oily plates. The oily surface is also uneven in thickness, and in many places may be entirely worn away by manipulation. The reason dipping in hot oil is considered to be effective is, that it is believed the hot oil penetrates every pore in the metal, thus displacing adhering dampness; but, it may be remarked, it would be preferable if it absorbed all dampness and yet became as dry and hard as cement. It is also claimed that the boiled oil adheres to the metal so perfectly that no exposure to air or moisture will make it separate. Such results can only be attained with the purest oils, with a surface that has dried, and under the most favourable conditions. To consider these are always to be obtained is a somewhat sanguine view to adopt. Reference to Chapter VI. will show that oils are not necessarily anti-corrosive. The objection to dipping plates, &c., as they leave the rolls, and before further manipulation, is sometimes obviated by brushing them over with boiling hot oil after handling, but this process is not to be compared to a removal of the scale on wrought iron or steel; and in the latter case it is especially necessary. Allowing wrought iron to corrode, then scraping off the rust, and immediately painting the cleaned surface, is another method adopted. If the systems mentioned are not used and the scale surface is coated, when it has to be repainted, the first coat of paint, or what remains of it, together with the scale on the surface to which it should

s 2

be found to adhere, should be scraped away with the paint, so that the pure iron can be painted for the *first* time, although its black scale was by the first coat of paint. Wire brushes, scrapers, sandstone, and the sand blast are used for removing rust. Steel-wire brooms are found to be more effective than coir brooms. Other methods adopted for removing fouling and rust are described in the subsequent pages of this chapter as more especially used for cleaning the bottoms of ships ; however, they can be employed or adopted for other structures, if considered desirable.

In order to prevent corrosion, it is of paramount importance that no water should be imprisoned underneath the paint. To seal up moisture by a coating of anything is to confine a corrosive agent on the surface which a paint is supposed to protect. To prevent such a confinement of moisture or incipient oxidation on the surface of iron or steel, the smaller parts of engineering structures, especially when they are to be shipped, have been heated to a low temperature by passing them through a furnace or otherwise, then they have been cleaned and coated with hot tar, if not required to be painted in colours, but, if they must be, linseed oil is substituted for the tar, and zinc or other paint used which will stand tropical climates.

If the painting has to be done by foreign workmen, care should be exercised to ascertain the habits, character, customs of the country, and to discover whether men are used to painting in an efficient and approved manner, and to know that the surface is properly prepared and cleaned before painting, the paint sufficiently stirred and amalgamated before being used, and that it is evenly and effectively spread with clean brushes. In some countries, Holland, for instance, much care is taken ; in some others, painting is effected in a haphazard manner, either from want of knowledge, carelessness, or from the idea that so long as the metal is covered with something, and the coat shines on application, it does not much matter. In the case of paint, and also any other substance, unless it is properly used, according to instructions prepared by skilled practitioners, it must fail or only partly succeed. The causes of unsatisfactory results with some paints are that it is not properly stirred or incorporated before being used; or the can, drum, or cask containing the paint is left open, and the nature of the paint so destroyed or impaired; or it has been made so long before being used that it has lost, partly or wholly, its preservative powers; or a clean brush is not employed in spreading it; or it is applied in unsuitable weather, or by unskilled labour. If oil paint has to be kept for some time in a can, &c., from which any quantity has been taken, by gently covering the top of the paint with water it prevents evaporation of the oil, and when the can is again opened, the covering of water can be poured away and the paint will generally not be too hard or thick, and will only require to be properly stirred before being used.

Painting iron without scraping or removing by acids, or other efficient process, the oxidation that has taken place, results, after a little time, in the outer skin becoming loose, in blistering, or in the coat falling in scales. Until the properly cleaned bare iron is coated, it is not protected. All wrought-iron or steel structures require regular and efficient attention by scraping away any discoloured places that may appear from time to time, and by repainting the metal, or the rust spots will pit the metal deeper and deeper and increase in size, notwithstanding painting. If this is not done, corrosion will at length become so serious that means may have to be taken to strengthen the structure or renew it. Scraping before a coat of paint is applied should be considered as necessary for the preservation of the metal, or blistering of the paint and pitting of the iron may result. Paint can be removed by heat and scraping, or by paint dissolver, &c. Paint dissolvers are of various kinds; some will soon dissolve paint and allow it to be wiped away. Caustic alkalies generally form their chief ingredient. A caustic alkali thus made is used: a very strong solution of soda is first obtained by mixing soda with boiling water, a flowing mixture of soft soap being added. It is then made thicker by adding fresh slaked lime so that it is of the consistency of thick pea-soup. It is laid on the paint to be removed from the surface, the time required being from about half an hour to two hours according to the strength of the mixture, the dissolver being kept moist. The paint will then generally be sufficiently dissolved to be rubbed or gently scraped away, the time taken also depending upon the nature of the paint to be removed.

The surface to be painted should be even, and especial care taken to ascertain, if fresh paint must be put upon old, that the old is firmly adhering, is effectively protecting the plates from corrosion, and that the new paint will not deleteriously affect that already upon the metal, or have its valuable properties and drying qualities injured. Paint spread on fresh tar, or tar put on paint, often becomes soft, does not dry, and is easily rubbed away. In repainting metallic structures, the best method to adopt is to remove the old paint, together with any scales of the iron, as the pure iron, not the outer scale or skin, cannot be properly protected unless this is done.

The methods generally adopted to clean a ship's bottom *without* docking are: (1) To drag a scraper along the keel and over the sides as far as the water line—a rather rough and imperfect way of cleaning. (2) By divers with a scraping apparatus hung from the ship, or preferably, as such submarine scaffolding costs very little, from ladders attached vertically to a lighter or punt, one being moored on each side close to the ship, and joined together underneath by horizontal ladders, making a subaqueous platform upon which the divers can work, the lighters or punts being moved as desired, and all parts of a ship's bottom thus be reached. (3) By a screen so constructed as to exclude the water from a portion of the ship's bottom, and enable a man in the dry chamber to inspect and scrape the plates. An apparatus on this principle was used on the celebrated *Great*

Eastern steamship. The first two systems of cleaning *without* docking are by far the most generally used.

In *dry* dock, scraping and cleaning can be effected by any of the systems used for unsubmerged structures if time allows, but this is not usually the case with the mercantile marine. Rafts are employed in the dock, often four, upon which men can stand to scrape and scrub a vessel's hull, as the water descends immediately pumping is commenced. The rafts and number of men should be so arranged that the cleaning proceeds *pari passu* as the water becomes lower, or the pumps may have to be eased or stopped; but if a ship's bottom is very foul it may be necessary to do so, or hang staging from the ship. Rotary wire brushes, scrapers, &c., are used to clean the surface ready for painting, and also fine ashes and sand with steam or water from a jet delivered at a high pressure, say about 120 lbs. per square inch, low pressure being of little or no use; the force required being governed by the magnitude and persistency of the fouling, the latter methods generally thoroughly effecting the scouring and cleaning.

Recently a machine to paint *without* brushes has been introduced, and has been used for ship and general painting. The paint is applied through a spraying nozzle by means of compressed air. The apparatus consists of an air pump, paint container, and spraying nozzle. It is said it succeeds admirably, reduces the quantity of paint required, produces very rapidly an excellent and uniform surface, and that paint so applied enters interstices and places which a brush cannot reach. Questions to be considered in painting by impulsion are : Will such atomic action separate the heavier from the lighter particles of the paint? Will paint so applied air- and water-seal a surface, and not introduce air, and force moisture before it, into the metal? Although a surface so spread appears to be even and continuous, is it really a film consisting of minute lumps and depressions?

A rotary air-compressor, driven by an electric motor, was used to mechanically paint some of the Chicago Exhibition Buildings. The air was compressed 15 to 18 lbs. per square inch, mixed with the paint, and the mixture was impelled through a hose from a nozzle in the hands of men on a scaffolding.

Occasionally it is necessary to artificially dry the bottom of a ship before painting. It can be done by portable fires, but care is required to prevent local heating, and for this purpose gas fires with atmospheric burners are to be preferred, as they are handier to move. It should always be remembered that the object is to evenly dry the plates, and not to heat them more than is absolutely necessary to remove all dampness.

With regard to the best time of day for painting operations, in fine weather about midday the atmosphere may be considered as being in its driest state, and about midnight in its most moist condition. As between midnight and sunrise the greatest deposition of dew occurs, outdoor painting should not be done till all the dew has evaporated or been artificially removed ; and any bare metal or surface should not be exposed more than necessary at any hour. Painting should not be done in rainy, misty, or frosty weather, or when the metallic surface sweats, as it often does, for instance, when the water-ballast tanks of a ship are filled, for then the plates are frequently colder than the air. The surface should be as dry as possible, and if a paint, without deterioration, possesses the property of suction and taking up any water that may be on the surface or in the pores of the outer metal, it is a valuable protection against corrosion, for then there will be no corrosion proceeding *under* the paint, as is not unfrequently the case when the surface of the paint appears perfectly sound. To cover up dampness is simply to invite corrosion. If a metallic surface be painted in frosty weather or when it has ice upon it, however thin—and it is often imperceptible to the eye, although upon the metal or in the pores—the result usually is that on the surface becoming thawed the paint loses its proper adherence, blisters, and finally falls off, for the frozen surface then becomes water, and no paint will adhere to water. It is necessary to do varnishing in dry weather to produce a gloss, as it can only be obtained by the rapid drying and evaporation of the oil and turpentine leaving, as it were, the gummy covering. Almost all paints will have a more glossy appearance if spread in dry weather than if applied in damp air.

It is advisable to consider the causes of the blistering of paint, which generally ends in its flaking, and in the exposure of the plate to atmospheric influences resulting in corrosion. It sometimes occurs from carelessness and over-haste in painting, which are fatal to durability; from want of preparation of the surface to be coated, which cannot be too clean or dry in order that all dirt and moisture are removed and air expelled; and also from the paint not being thoroughly mixed or equally spread. Blistering in ship paint may only extend to the anti-fouling or *second* coat, or to the *first* or anti-corrosive coat. In paints generally, it principally arises from the second coat being applied before the first has had time to dry and harden; from imperfect incorporation of the ingredients, or mixing of the paint, with consequent inequality in its texture and general properties; the use of driers that injure it; over-oiling of the paint, i. e., too much oil being present in the second coat, a larger quantity of turpentine being desirable in each succeeding coat in ordinary oil paints; by moisture having accumulated at some crack or joint, resulting in a thin layer of rust between the paint and the surface of the metal; the evaporation of moisture confined under the coat; by heat; by air being confined upon the surface consequent upon the too rapid and careless spreading of the paint; by inequalities and unevenness of the surface painted; and also from decomposition of the paint.

For moisture to remain under the paint, the following conditions may be said to be necessary. That it is upon the plate when the paint is applied, and is confined by the coat of paint not possessing the power of suction. The degree of blistering is regulated by more moisture being present at one place than another, and by the irregular evaporation of liquids contained in the paint, which, in some measure, accounts for blisters frequently being raised in the centre, and only being attached to the metal or plate at the edges.

Blisters formed in the second coat, and upon the first, may also arise from the paint underneath not having been thoroughly dry, and it usually happens that a large surface does not dry equally in time or manner. The boiled oil in oil paint contains moisture, and, in the case of linseed oil, if steam is used to obtain the oil from the seed, the quantity of moisture is increased. Again, turpentine or a volatile liquid is present in most paint, or in the driers that are frequently mixed with it; and in the employment of volatile liquids in paint, an evaporating mixture, subject to increased action by heat, &c., is more or less formed, which possesses a tendency to reach the air. When the coat of paint is not too hard, and possesses sufficient extensibility, the surface is raised usually in a more or less round spot, called a blister, having a centre of energy ; but when the paint is sufficiently hard and inelastic to resist such local distortion, it cracks, the result in each case being exposure of the plate, and corrosion. Consideration of these operations indicates the importance of the surface of the plate being thoroughly dry. This may be attained in the case of painting other structures than ships ; but the latter must frequently be coated in variable weather, such as mist, slight rain, frosty air, and under all other conditions of moisture known to be deleterious to metals, because of the requirements of commerce; however, the surface can be artificially dried as has been just previously described. Therefore any composition or paint that absorbs any moisture on the plate, *without deterioration of the first or other coat of paint*, possesses a valuable anti-corrosive property, and one especially to be desired in a ship or any other paint.

There is another kind of blistering, the result of one coat

being placed upon another. It proceeds from the rapid absorption by the first coat of moisture contained in the second coat, thereby destroying or deteriorating the powers of adhesion and homogeneousness of the second coat of paint, and producing what is called dry-blistering. Want of thorough and equable incorporation of the ingredients of which the paint is composed, and uniformity in the drying qualities of the composition, and a surface being painted in dry and in damp weather, or in the sun's rays and in shade, or before the sun had dried over the *whole* area any dew or moisture that may be present, are the principal causes of it, and are generally to be prevented by care. However, it is not easy to obtain a uniformly firm adhesion between paint and the surface to be coated. A smooth, polished surface, with ordinary care, can be so painted that the adhesion is firm, equal, and permanent, until the virtue of the paint is exhausted ; but on a rough surface, such as a casting, or a structure having many abrupt bends, sharp angles, edges, recesses, external flanges, tie-bolts, or anything to interfere with easy and equal spreading of the paint, it is somewhat difficult to attain. In using oil paints on rough surfaces, or such as are not deleteriously affected by the oil, metal has been first washed with hot oil, the idea being to penetrate the pores of the metal to be coated, remove any dampness that may be upon it, to make an air-tight and waterproof coat, and one to which the paint will permanently adhere. Whether, so long as this hot oil first coat retains its desired properties, corrosion is prevented, depends much on the quality of the oil used, the evenness of the coat, and the cleanness of the surface upon which it is applied ; but it decreases the chances of blistering by hindering the production of rust and consequent blistering or flaking of the paint.

CHAPTER VI.

NOTES ON SOME OILS, GUMS, AND RESINS USED IN PAINT.

BEFORE considering anti-corrosive and anti-fouling paints, it may be useful to refer to some of the oils, gums, and resins used in paint, with the object of indicating some of the salient characteristics so far as regards their anti-corrosive or corrosive effect, durability, and other desired properties, and *not* with the view to describe or explain the method of manufacturing paints, compositions, or varnishes.

The varieties of oils are very numerous, some are fluid, such as olive, linseed; but the semi-fluid, i. e., those which require a higher temperature than that of the air to make them fluid, closely resemble the fat of animals. Oils and tallow are principally composed of carbon and hydrogen gas, the solid part being carbon, and the volatile, hydrogen gas. In using oils in making paint, it is well to bear in mind that some quickly oxidise on exposure to the atmosphere, and form a thick gummy mass, and that generally, as soon as the oil is thus gradually removed, the paint or varnish becomes brittle, and begins to crack, then heat, frost, rain, and damp will soon wear away the coating. Some become resinous when exposed to a slight degree of heat a little above the temperature of the surrounding atmosphere. Oils when associated with alkalies become converted quickly or slowly into soaps soluble in water. Linseed oil is one of the most easily saponified, for if it be shaken with a solution of potash, soda, or ammonia, it unites with the alkali, forming a thick emulsion of soft soap. Ammonia is especially present in the air where organic decay is proceeding. Of course, in the immediate neighbourhood of manure, or heaps of decayed vegetable matter, saponification to some extent will follow. When dust impregnated with ammonia has been found on varnish on carriages, &c., in stables, the varnish cracks, usually first on the upper surfaces and projecting parts, therefore, as moisture is usually present, it is well to keep them covered. Ammoniacal vapours, whether they arise from manure, a stable, urinal, or from decaying vegetable matter, are active corrosive agents, and any substance exposed to them requires special protection, for it may become dissolved or cracked from being subject to such decaying influences.

Atmospheric action differs greatly both in nature and degree upon various oils; some become viscous, and increase in weight; and some, on the other hand, lose weight; others appear to be unaffected; but all, whether animal or vegetable, more or less absorb oxygen from the atmosphere. Drying oils, such as linseed, cotton seed, and a few of the vegetable oils, may become in a few hours, if exposed on a surface, quite gummy, and appear like a varnish. Mineral oils have not the affinity for oxygen that the others possess. The sun's rays and heat have an oxidising and extractive influence upon oil, whether in a film of paint or in the pores of a metal. They attract it to the surface, and then the lighter portion evaporates, and the heavier congeals, deterioration is decided, air and moisture find their way to the metal through cracks or hair-like openings in the paint, it loses its adherence to the metal, and only acts as a cloak to hide corrosion proceeding underneath it.

All oils, resins, and gums exposed to air, and especially to the light of the sun, oxidise and burn away with more or less rapidity, leaving a powdery residue behind. Lead and zinc paints both depend for adhesion upon the oil or varnish with which they are mixed, therefore, when the oil or varnish in such paint perishes, the paint, or rather the residue, falls off, not because of the action of water, but because of the effect of light and air. Gutta-percha, by way of illustration, becomes rapidly deteriorated when exposed to light and air, but water does not appear to affect it. Dampness makes them soluble, for they are hygrometric, notwithstanding that they may first dry and harden.

In *Iron*, of the 31st March, 1893, in an article by the author of this book, the following paragraph occurs:—

"Although oil-paints are well adapted for coating wood, for metallic surfaces they are not so suitable; however, wood-tar is more penetrating, and possesses antiseptic properties which oil-paints do not, and is cheaper, although dearer than coal-tar, which has not equal antiseptic or penetrating qualities. Raw linseed oil impregnates wood more than boiled oil, and forms a kind of resinous adjunct which is to be desired, but it does not do so in the case of metals; therefore, its chief object is to prevent moisture reaching the surface of the metal, for while moisture, i.e., a moderate degree of wetness, does not affect the oil, both light and air have a deteriorating effect upon it, and damp or moist air especially so, inasmuch as it becomes partly dissolved by it and assumes a liquid form, particularly when the air is charged with moisture and no evaporation takes place. The oil has another special purpose, viz., that of causing the pigments or other ingredients of the paint to adhere to the surface and to become a united mass or thick liquor. Hence the oil may be said to be by far the most important constituent of what are called oil-paints, and the importance of its purity is great; and, as it is liable to deterioration from such causes as those referred to, it can hardly be said that it is suitable for exposed work; and when applied to engineering metallic structures in the open air, its anti-corrosive and protective life is short, for, the damp air having dissolved it sufficiently, this action only being a question of time, however slow its progress may appear to be, moisture will permeate the coat of paint, or rain be forced in by wind, the consequence being that the metallic surface becomes corroded, and as rust forms, the paint either blisters or more frequently becomes detached. One reason why ordinary oil-paints are not well adapted for out-door ironwork is that the turpentine usually employed in making them is distilled from impure turpentine by the medium of water, the volatile oil being so collected and the resin remaining; it is therefore a volatile oil useful in the case of painting wood, but not so with comparatively non-absorbent substances, such as iron or steel. As a rule, the greater the quantity of turpentine in oil-paint, the more likely it is to have a dry scale on its outer surface."

Numerous experiments by Dr. Sace, of Neuenberg,* conducted with a view to ascertain the nature of different resins, showed that copal, amber, dammar, common resin, shellac, elemi, sandarach, mastic, and Caramba wax, can all be reduced to powder. That the following will become pasty before melting: amber, shellac, elemi, sandarach, and mastic; the others will become liquid at once. Dammar and amber do not dissolve in alcohol; copal becomes pasty; elemi and Caramba wax dissolve with difficulty; while resin, shellac, sandarach, and mastic dissolve easily. Caustic soda dissolves shellac readily, resin partly, but has no influence on the others. Oil of turpentine dissolves neither amber nor shellac, but swells copal, dissolves dammar, resin, elemi, sandarach, and Caramba wax easily, and mastic very easily.

Boiling linseed oil has no effect on copal, amber, and Caramba wax; shellac, elemi, and sandarach dissolve slowly in it; while dammar, resin, and mastic dissolve easily. Petroleum ether has no effect on copal, amber, and shellac; it is a poor solvent for resin, elemi, sandarach, and Caramba wax, and a good one for dammar and mastic. Benzol does not dissolve copal, amber, and shellac, but does elemi and sandarach to a limited extent, and Caramba wax more easily; while dammar, resin, and mastic offer no difficulty. It is advisable to remember these, and the subsequently-mentioned experiments in selecting gums and resins for paint, or paint made with them.

Gums and resins are generally distinguished from each

* See Dingler's Journal.

other by the solubility of gums and insolubility of resins in water, and the solubility of resins and insolubility of gums in alcohol. Some are partially soluble in both solvents. They are then called gum-resins. Turpentine dissolves many resins, and is also useful in mixing and drying paints. With regard to gutta-percha, i.e., the gum of the percha, and its use for cable preservation, some gutta-percha contains more resins than others. The resins are considered to increase the insulation, but facilitate the deterioration of the gutta-percha, and although different kinds of gutta-percha may at first possess the same properties of insulation, that having the *least* water and resin in it would remain the longest in good condition. Chemical analysis of gutta-percha is, therefore, necessary in order to ascertain the impurities, as they may be called, in it. If there is a complete absence of water, it adds to the durability of the material and increases the insulating power.

Some experiments show that cotton-waste, soaked in linseed-oil varnish and exposed to the atmosphere, becomes gradually heated, and in five hours was almost ready to burst into flame. Dr. Rohrig observed that this property of waste soaked in varnish has long been recognised. Oils in general are known to absorb oxygen when in contact with the atmosphere. Some oils become thereby solidified and hardened, while others retain their liquid condition : the first mentioned being used for varnishes, the latter for lubricants. In the former case, a gaseous body is transformed into a solid ; in the latter, into a liquid condition ; and it is known that neither of these processes can take place without the evolution of heat. By being soaked into the cotton waste in a very finely-divided condition they are exposed to the action of the atmosphere over a large extent of surface, and therefore action is much more rapid and intense. Dry varnish is thus varnish oil, saturated with oxygen ; while liquid varnish, as prepared for use, is varnish oil which has absorbed oxygen, but not enough to saturate it. Such varnish is prepared by mixing with the oil some sub-

stance containing oxygen, and then either leaving the
mixture in a warm place to complete its union, or heating it
at once to boiling point. The latter process causes the oil
to take up a larger proportion of oxygen than the former,
and probably the proportion which still remains to be absorbed before it solidifies is too small to heat the waste to
combustion point. It would, therefore, seem desirable to use
the latter process, and also to make a minute analysis of
varnishes, &c., so as to determine under what conditions
they may be used with safety.

Some experiments, which were continued over a year, were
recently made in the United States of America to ascertain
the effects of oils in contact with various metals with the
following results. Iron was least attacked by seal oil, and
most affected by tallow oil. Zinc seemed to be but little
attacked by mineral lubricating oils. The least action was
with lard oil, and the greatest with spermaceti oil. Copper
was not attacked by mineral lubricating oils, and was least
affected by spermaceti oil, and most attacked by tallow oil.
The following table gives a *résumé* of the experiments.

Oils.	Metals not attacked.	Metals least attacked.	Metals most attacked.
Mineral.	Zinc and copper.	Bronze.	Lead.
Olive.	..	Tin.	Copper.
Colza.	Bronze and tin.	Iron.	Copper.
Tallow.	..	Tin.	Copper.
Lard.	..	Zinc.	Copper.
Cotton Seed.	..	Lead.	Tin.
Spermaceti.	..	Bronze.	Zinc.
Whale.	Tin.	Bronze.	Lead.
Seal.	..	Bronze.	Copper.

The metals to be tested were first thoroughly cleaned
with ether and dried. The weights were then taken, and
the metals placed in closed tubes filled with the oils, and
kept for 12 months at a mean temperature of 27° C. during
the summer and 10° to 15° C. during the winter, corresponding

T

to about 80° F. and 50° to 59° F. respectively. These experiments are here referred to in order to illustrate the varying effects of many oils on metals and to show that *all* oils do not protect metals. It is well known that the quality of the oil used as a lubricant, which should be unctuous, free from acid, and non-drying, much affects any destructive effects on a boiler. When free fatty acids are present, they are, as it were, neutralised at the expense of the metal, but chemically neutral oils do not generate deposits nearly so much, and some experiments seem to show not at all, if not employed in excess.

Linseed oil is generally considered the best for mixing with colours or varnish, and when spread with a brush on a sheet of glass it should dry in about two to four days. If some 20 per cent. of sulphuric acid is mixed with drying oils it causes a quick rise in the temperature. Oils are sometimes agitated with a small quantity of sulphuric acid in order to get rid of impurities. If this acid is not subsequently removed it has a corrosive effect on iron or steel, and also acts prejudicially on the oils. A method used to remove the acid is by mixing and agitating the oil with an aqueous solution of chloride of barium, or some alkali. When left to settle, the acid is absorbed by the water, sulphate of barium is precipitated, and the oil floats above the water with a brighter and clearer colour than before. Any oil containing sulphuric acid, such as cheap kerosene oils, should not be allowed to be spread upon metalwork. Linseed oil is known by its great density and medicinal smell, and by agitating it with sulphuric acid, when, even if there is only a small quantity of it present, the oil will be converted into a brown paste.

It should be ascertained that the oil used to make any paint for metalwork does *not* contain any acids, for many oils do, and, therefore, instead of being *anti*-corrosive, may be actively corrosive. This may be overlooked in what is euphemistically called "good oil paint," which may have anything but "good oil" in it for anti-corrosive purposes. Many refined oils made from animal and vegetable substances contain a very appreciable percentage of free fatty acid which has a corrosive action on metals, and in their natural state nearly all animal and vegetable oils contain such acid, increasing with age, &c. When the oil is treated with an alkali the free fatty acids are converted into a soapy mass. If a fatty oil has to be used on machinery it should be neutral, or entirely free from acids. Mineral hydro-carbon oils have not the chemical action on metals that fatty oils have, for the free fatty acids are absent. The acids are liberated from fats and oils in the process of manufacture by steam at high pressure in the presence of water, sometimes with a little lime added. One of the chief requirements of an oil for anti-corrosive purposes is that it should be perfectly neutral, and develop no acid when mixed with other materials used in any paint. Raw linseed oil being usually used for white lead colours, they cannot be said to be free from corrosive action on metals if there is any acid in the oil. Priming for inside work often consists of white lead, raw linseed oil, red lead, and driers; and for outside work, boiled oil is used instead of raw linseed oil. For common white paint, white lead, driers, linseed oil and turpentine are used, raw linseed oil for white colours, and boiled oil for tinted or dark colours, because "boiled oil" causes white lead to assume a yellowish tinge, hence *raw* linseed oil is often employed in preference, also because boiled oil somewhat lessens its covering properties and the body of the white lead. Boiled oil is sometimes made of 1 gallon of raw linseed oil, 6 ounces of litharge, added to make it dry quickly, 3 ounces of red lead, the whole being allowed to simmer for about a couple of hours. It is then sufficiently defecated to be drawn off for use in about ten to twelve hours.

Oils should have no acids in them, whether fatty acids from fermentation of the oil itself, or from the process of refining, be free from foreign matters, be reasonably fluid, and not become gummy through oxidation. Mr. Watson's table shows the amount of metal dissolved by the oils mentioned after ten days' exposure to copper, and twenty-four days' exposure to iron.*

* See 'Industries,' July 23, 1886.

T 2

	Copper, grains.		Iron, grains.
Linseed oil	0·3000	..	0·0050
Olive oil	0·2200	..	0·0062
Almond oil	0·1030	..	0·0040
Colza oil	0·0170	..	0·0800
Castor oil	—	..	0·0048

By some, this corrosive effect is not attributed to the oil, but to the presence of mineral acid used in refining the oils, or to free fatty acids engendered by fermentation in the oils. Mineral oils have an especial value because of their slight tendency to oxidise when exposed to atmospheric influences, therefore they have been used for protecting bright metal surfaces, and are so employed in various climates for military purposes, for mineral oil does not contain the acid that vegetable and animal oils possess. The liquid products of paraffin oil, or rather, the destructive distillation of shale, are known chemically as hydro-carbons. They are separated into groups, the group of the heaviest gravity being known, as stated by Mr. Coleman, as lubricating oils, those of medium gravity being used for cleaning machinery, and those of still lighter gravity being employed for burning purposes; the brightest being known as naphthas and gasoline.

Vegetable gums and oils that are insoluble in water, and dry hard paints and varnish, become soluble and liquid when dampness is accompanied by mould; also oil, resins, and gums exposed to the air and sun oxidise and evaporate. Oils and varnishes are especially soluble in damp places, and oils have less power of resistance against salt water than gums. Although oil and water cannot by agitation be made to unite, they will do so upon a solution of caustic potassa being added, the alkali acting as an intermedium. This action is explained by the supposition that the oil and alkali form, in the first instance, a compound which is soluble in water. A peculiar property of oil is that it will permeate what may be otherwise regarded as a sealed and water-tight substance.

CHAPTER VII.

NOTES ON LEAD AND CHEAP PAINTS, AND THE REQUIREMENTS OF AN ANTI-CORROSIVE PAINT.

WITH regard to anti-corrosive paints, the durability of iron or steel structures can only be secured by protecting them from rain, vapours, acids, salts, and all other corrosive influences, for unless they are properly painted with a really anti-corrosive coating and carefully inspected at reasonable intervals they will be subject to corrosion. The regular inspection and repainting is of great importance, and unless it is properly effected, the most carefully designed and magnificent structure will deteriorate and gradually lose much of its strength by corrosion. To prevent corrosion every agent promoting decay or decomposition must be excluded from the metal, and it is important to remember that any preservative coating in many cases can only be periodically renewed. Complete preservation from oxidation in bridges, roofs, piers, &c., is almost impossible to attain; but if every care be taken to extract any moisture before covering the metal, to procure the best anti-corrosive and insulating paint, to maintain the coating, and to remove all films of rust that may appear here and there in spite of all practicable care, the serviceable life of metallic structures, so far as regards decay from corrosion, may be expected to be very greatly prolonged. The result of due care has been shown in the complete preservation of the renowned Britannia tubular bridge, erected in 1848; however, *un*protected iron of the dimensions usually employed in construction will corrode in a few years, even under most favourable circumstances, and its serviceable life may be but a few months under *un*favourable conditions.

Chemical action is not desirable in a paint for metals, unless it can be proved that it does not in any way affect their strength or durability. A physical, physiological, and toxical action is to be desired in order to prevent corrosion and fouling by vegetable or animal life or micro-organisms, but the antiseptic power should be as permanent as possible, and therefore its volatilisation and solubility in water should be prevented as much as practicable, or the utility of the antiseptic and anti-corrosive agents is destroyed; in fact, durability and insolubility are of the greatest importance. An effective anti-corrosive and anti-fouling paint may be said to have a three-fold action, the physiological and toxical action of making a poisonous or distasteful covering to every kind of life so as to prevent its attachment. The chemical action of drying, hardening, and adhesion to the metal, yet not to such a degree as to cause cracking, chipping, or to destroy the required elasticity of the paint, but yet sufficient to make a moisture- and air-tight covering. The mechanical action of penetration to and filling the pores of the metal, and the all-important one of sealing the surfaces so that no air and moisture can reach them. The mechanical action of an anti-corrosive paint in closing the pores of the metal is most valuable, as it prevents any subsequent absorption of moisture. If the surface be sealed from air, moisture, and the usual corrosive agents, and any dampness upon the bare plate be absorbed by the anti-corrosive paint, then it may be said almost all has been done that can be done by paint to prevent corrosion, and if any occurs it would appear, so far as present knowledge enables a judgment to be forming it can only be caused by moisture, or other corrosive influence that has previously permeated the pores of the metal, forming a nucleus of active corrosion aided by galvanic action brought about by inequalities in the texture and composition of the metal.

It is sometimes thought all that is required in any paint

is to keep the humidity of the air from the metal, but there
are other objects, which will be stated ; however, the extraction of any moisture that may be on the surface when it is
painted is a most important one, for if water or moisture is
confined in the pores or on the surface of the metal, corrosion
is invited ; hence oil paints and substances having no affinity
for water are not to be particularly desired for covering
metals, but those which have some power of absorbing any
moisture on or near the surface, *and then become dry and
elastically hard and finally set*, somewhat after the manner of
Portland cement, although possessing greater adhesion and
elasticity.   All paint used in bridges or structures subject to
vibration should be of such consistency as to be elastic, or it
will crack or become detached.   It should always be
remembered, in choosing a paint, that a metal surface is
unlike a wooden one.   The former is not absorbent, and the
paint has to be so made that it will adhere to it.   Timber,
however, to a certain extent, absorbs or attracts the paint,
therefore, a paint that may be excellent for wood may be of
little or no use for metals.

There is one point which it is well to always bear in
mind, namely, that although metals may be coated with
paint or other protective substance, no paint can be indestructible in the full meaning of the word, for in all, by
the laws of nature, decay commences almost from its application, depending upon the purity, quality, and behaviour of
the material used, and the degree of intensity of the corrosive
and all other decaying influences.   Periodical applications of
any preservative coating may after erection be prevented by
constructional arrangements ; however, to consider the initial
painting sufficient is but a comfortable way of leaving the
question of preservation to be solved by others.   The chief
objects in a paint being to extract any dampness from the
metal and to prevent air and moisture reaching it, oxides and
carbonates of lead, &c., are not very suitable for pigments for
anti-corrosive paint, for they may absorb air, and, when they
have thrown off the oil, may convey it to the iron.   As soon

as the oil in oil-paints has been exhausted the film of paint becomes a kind of porous plaster. Mr. E. Guber, Mem. American Society C.E., has recently * examined a large number of iron and steel bridges, with a view to ascertain the best means of preserving such structures from rust. *In all cases rust was found beneath the paint.* In some instances the amount was very little, but in others the condition was much worse. In most cases the rust occurred in spots. The smoother the surface the less the rust, plates being generally much freer than angle-irons and similar sections, and eye-bars were also usually clean. Parts that had been heated in the process of manufacture also showed less rust than others. In cases in which the metal had been oiled in the shops, there was much less rust under the paint than when this precaution had been omitted. As the rust spots under the paint were no worse in the old bridges than in the new ones, it was concluded that the corrosion was not proceeding actively. Iron oxide paint seemed to get tougher and more adherent with age, save where the coating was very thick, as in cases in which the repainting had been done at short intervals. It was then comparatively soft, and could be easily removed. In the case of bridges which had been painted with red lead, the paint was universally brittle and very easily removed; and in that of one bridge, painted with white lead as a finishing coat, the paint was cracking wherever exposed to the sun.

As all paint spread upon ironwork should not contain anything injurious to the metal, it would be well if it were analysed before being used. Naturally, no manufacturer of paints would desire to give a full analysis of the composition of a paint. To meet this reasonable demur, a copy of the certificate of an analytical chemist might be presented with it, stating that no substance used in the paint had a corrosive effect on iron or steel. This would eliminate any paints that may be misnamed anti-corrosive paints. In adopting metallic paints the effects on the iron of the different metals

---

* See 'Engineering,' May 3, 1895.

from which they are partly derived should be considered. It is sometimes expressly stipulated that mineral colours only, without any admixture of lead or zinc, are to be used in the painting of railway iron roofs, on account of the sulphurous fumes from the locomotives.

In a damp situation, vegetable gums and oils, paints and varnishes, usually become soft and more or less liquid, but poisons which are destructive to microscopic plants, mildew, mould, &c., and all other life promoting decay, prevent or lessen decomposition until they lose their power. It is well to remember, in considering anti-corrosive paints, that in a chemical union the properties of the mingled bodies are altered; in a mechanical admixture they are not, and, therefore, can be again separated. The importance of a chemical union of the constituents of a paint is clear, for if the ingredients can separate they cannot form an even covering or be a water and air-tight coat. The *uniform* efficiency of a preservative coat will, in great measure, determine the freedom from corrosion of a surface covered with it.

It is very probable, when the corrosion of many existing structures has become more apparent, it will be expressly stipulated that no *white lead* shall be used in any paint for any metallic structure of the slightest importance, but particularly in the first coat, viz., that to be spread upon the bare metal. Zinc-white, which is a powerful anti-fouler, and does not affect oil, or red lead should be used in preference, and never any but the best and purest to be obtained in a paint used for structures of importance, for cheap paint is only a cloak to hide decay, and may be composed of the most inferior materials, selected without a care as to whether they have anti-corrosive properties or not, and be simply mixed so as to be glossy, and to shine at all hazards. Paint that is twice the *first* cost of any specious stuff sold at prices below those for which even the *raw* materials of a really anti-corrosive and efficient paint can be obtained, is much the cheaper in the end, for cheap paint may soon lose any protective qualities it may at first possess, and frequent repainting is necessary.   On the other hand, in the case of a pro)perly-constituted and made paint, the protection is effectiive, no oxidation is proceeding *underneath* the paint, and repauinting is only required at much greater intervals, and then probbably, with the exception of a few places, only *one* coat may be required.

If red-lead paints are used, it is important to know that red lead is employed, as it is comparatively expensivve, and coloured earths may be substituted.   When mixed wiith the commonest oil, the paint may be called good lead paimt, and be thought to be anti-corrosive, whereas it is really thte very reverse, for there is a chemical action of lead upon iroon, and if it be mixed with common oils having acids in them it is objectionable, and should not be called anti-corrosivce.   In the case of the bridges mentioned on page 280, it was found that the red-lead paint was universally brittle andl very easily removed, probably owing to the evaporation or (oxida-tion of the oil.   The white-lead paint was also craacking wherever exposed to the sun.   Oil paints do not long : retain their preservative and anti-corrosive qualities, partly beecause the moisture of the condensed aqueous vapour on thte sur-face coated gradually causes the permeation of the paint. Then corrosive action commences, and rusting occurs : under the paint, and the latter becomes detached in time.   The evaporation of fluids being in proportion to the siurface exposed, and not to the mass, paints are severely trieod, for they may be subject to the sun's rays, wind, warm aiir, and other volatilising influences, on being applied. Some paint also contains a percentage of water, which may vary/ from one to two per cent. or more.   It is a decided advantcage if it contained none, but received its necessary fluidiltty by means of non-corrosive agents only.

The preservative value of paint or composition canmot be otherwise than dependent upon the quality of the ingrecdients used, and a really good oil paint cannot be made unlesss the materials are of the best and purest description, especiallly the oil and red or white lead or zinc.   If cheap white leaad is used, the paint will have a yellowish tinge, and will probably chip and scale. The whole of it must be finely and evenly ground and thoroughly mixed. Oil paint made under such conditions cannot be supplied at a reasonable profit at the prices at which what is called "good oil paint" has been sold. Comparatively few paint and colour retailers know much about the anti-corrosive and anti-fouling qualities of the ingredients they buy to sell, which, however, may or may not be good, for the latter is a conveniently elastic word, for they buy the ingredients with the laudable object of profitable trade. It is the *buyer* who is to blame, and who deceives himself, and not the seller. First cost, unfortunately, is considered by many as everything, the quality, durability, anti-corrosive, and preservative and anti-fouling properties being disregarded. There can be no question that the purity of any oil used for mixing is of paramount importance, and no good or durable oil paint can be made without using the best for the purpose, and unless the pigments are of excellent quality the colours will fade rapidly when exposed to the atmosphere, rain, and the sun's rays.

White lead has been used in paint for a very long time, but white oxide of zinc, which is more waterproof than lead, only comparatively recently. Dry ground white lead requires a considerably less quantity of oil to moisten it than an equal quantity of zinc, but the latter is the more durable and has other advantages. A surface coated with white lead and oil paint, and exposed to the sun's rays and general weather influences, will, after a time, most frequently be found to be cracked, and microscopical examination will show other defects. It is well to remember that the lead particles in paint do not possess cohesiveness, the oil imparting it to them; however, oil does not incorporate well with most pigments. Lead paints, even if they are to be preferred for any work, most certainly are *not* for any that is submerged, for water removes the cohesiveness given to lead paints by the oil. The sun's rays also cause oxidation and deterioration. Zinc, on the other hand, forms a continuous covering, as the particles possess considerable cohesion, and do not separate nearly so easily, but appear to hold together on metal much better. It is also not so affected by the sun's rays, rain, and air, and the surface is smoother and harder, and therefore water and dirt do not remain on its surface so readily, nor is it so *unevenly* affected.

White-lead paint, when exposed to the sun and atmospheric influences, loses by degrees the oil which acts so importantly in binding the particles together, and then the oil lead paint becomes a lead composition having but little cohesion, and partakes of the character of a whitewash, and is removed by the joint action of rains, wind, and air, and becomes a kind of powder. The time is probably quickly coming when, except for indoor or the commonest purposes, white-lead oil paint, i.e., white lead ground in oil, diluted with perhaps some linseed oil, turpentine, and a little driers, added to make it dry quickly, will cease to be used for any engineering structure, except of the meanest kind, and its use simply be confined to domestic purposes and to coat things of little value, but which must be covered with something or other for sake of appearances. Common white lead, the poisonous carbonate, not only is injurious to those who make it, but is the cause of disease to those who use it in paint, as witness painters' colic, paralysis, &c. As it is by no means an efficient preventive of the corrosion of metals, it is difficult to understand the affection that some appear to have for it. At present, white lead is the basis of most ordinary paints, and it has been estimated forms about $\frac{9}{10}$ of their composition. The ordinary and old-fashioned lead and oil paints are very seldom indeed now used for the interior surfaces of iron or steel ships, some effective anti-corrosive composition being employed for that purpose.

At a recent meeting of the Association of Engineers of Virginia, Mr. Wallis read a paper on painting ironwork,* in which he recommended that the first coat should be red lead ground in raw linseed oil, used within two or three weeks

* See 'Engineering,' June 8, 1894.

after mixing, and kept thoroughly mixed when in use. This coat would dry in from 24 to 30 hours. For a black finish the next two coats should be made up from a paste composed of 65 per cent. of pigment and 35 per cent. of raw oil. The pigment should consist of 65 per cent. of sulphate of lime and 30 per cent. of lamp-black, to which should be added 5 per cent. of red lead as a drier, the whole thinned to a proper consistency with pure boiled oil. For a red or brown finish the paste should contain 75 per cent. of pigment and 25 per cent. of pure raw oil. The pigment in this case would consist of 55 per cent. of sulphate of lime, 40 per cent. of oxide of iron, and 5 per cent. lime carbonate as a drier. Lead paints were not recommended for finishing coats on account of chalking, nor zinc paints on account of cracking. The general objections to lead and oil paints that have been stated herein apply to the first coat recommended.

Paints sold at a very low price have been made of Paris-white or any suitable and easily obtained earthy substance which will mix with the cheapest liquid that can be so made as to look of some consistency, take up colour, and appear glossy ; but a great deal more than a mere mordant is required in a really anti-corrosive paint. The best materials to make an anti-corrosive and anti-fouling paint cannot be obtained for much less than about 15s. to even 20s. per hundredweight of mixture in moderate quantities, *not* each ingredient, to which must be added the cost of mixing, evaporation, drumming or barrelling, labour, superintendence, rent, taxes, cost of delivery, traveller's commission, credit, bad debts, and all other usual business charges and expenses, such as advertising, to say nothing of profit. It will be judged that the prices at which very cheap paints have been sold do not admit of the best material or quality being used. Paints delivered free at 1½d. per lb., however sufficiently good for purposes supplied—and it is only fair to paint-manu-facturers to say they do not affirm such a mixture to be the best paint—cannot be regarded in the same light as that carefully prepared from the very best materials specially selected to preserve a metallic surface from corrosion and fouling. An idea of their possible lowest price may be gathered from what has been here stated, but, of course, the prices will vary according to the current prices of the raw materials, labour, &c.

If the different ingredients of which paint is made are just, as it were, poured or cast in any old pail or receptacle, and stirred up till they do not look streaky or unmixed, it cannot be a homogeneous mixture. As hand mixing is an unpleasant and laborious process, it may be stated that hand-mixed paint is not unlikely to be *not* properly incorporated, and it may be taken as an axiom that it will never suffer much from being *over*-mixed. It should be effected in a closed building in which the temperature is moderate, be kept from moisture, rain, the sun's rays, dirt, and any other deleterious influence. Those who are not conversant with the manufacture of paint it is well should know that for a paint to be really properly mixed and thoroughly incorporated, and not a mere haphazard mechanical mixture liable to become separated at any time, even on drying, great care is required both as to the preparation of the ingredients, the time necessary for amalgamation, degree of heat, complete solution, defecation and removal of scum, and various other operations, such as mixing the component parts by degrees, some at first separately, or, may be, in a different way. By care, any volatile liquid can be so employed by experienced manipulation that comparatively little is lost.

To a certain extent, it may be said that paints are not unlike some cements which have several ways of becoming hard, viz., by drying, congealing by oxidation, in a few cases, by cooling, and in others, by becoming set by chemical changes. Oils are said to dry, not because of evaporation, but by absorption of the oxygen from the air. These harden from the external surface inwards, at least the film is *supposed* to be so caused to be wholly hardened. Another process of hardening is that in which a substance, as cement, first forms a chemical combination with water, and then slowly or quickly hardens by drying, some of the water evaporating. It is well to remember these different processes in considering that of paint.

Homogeneousness of the paint and intimate contact between the paint and metal is necessary, for there is always a layer of air on all substances, and the endeavour should be to expel it in spreading the paint, for carelessness in painting may cause blisters by confining the air if the paint is simply daubed on.

Where ironwork is submerged or occasionally submerged, approved anti-corrosive and anti-fouling paint, such as is used for ships' bottoms, should be employed, as it is specially manufactured to resist the effects of water and fouling.

There is some secrecy as to the composition of " driers," but litharge, sugar of lead, copperas, white copperas, and terebene driers, are often used, copperas serving as a hardener. If different " driers " are employed, their action will be somewhat dissimilar. It is, therefore, advisable to use *one* kind for any paint for *one* structure. Some of the best paints are so constituted that none are required. A rough test to ascertain if a paint is likely to adhere properly and not scale off, is to dip a painted plate in cold and also in hot water, and then to test it with a hammer.

Paint used for painting ornamental ironwork should be of the best anti-corrosive character, and not more than two coats be applied, a thick layer spoiling the delicacy of outline. The surfaces must be properly preserved, for if they are corroded, the whole design becomes an eyesore instead of an attraction. Linseed oil of the best quality, applied when the metal is hot, or zinc paint, are suitable for such metalwork. Red, brown red, or purple-brown, and sometimes yellow ochre and grey to receive light coats, are the principal colours used in painting ironwork the first coat at the foundry. Ironwork for hot climates should be painted a light colour, so as to keep the metal so coated as cool as possible, and lessen expansion and contraction, which to some extent aids corrosion, for it strains the metal, tends to open the joints, makes the parts loose, tries, and deteriorates the paint. Structures are painted white so as to increase the radiation and diminish the effects caused by alterations of temperature.

In the antiseptic treatment of wood, the chief aim is to fill and close the pores of the wood, and air- and water-seal the surface, and, as far as practicable, it is so in the case of metals. It may be, therefore, advisable to call to remembrance that experience with impregnating preservatives shows that those which penetrate most easily usually retain their properties for the least time. The preservative required is one that is the most efficient and most permanent. In selecting an anti-corrosive paint, its durability has to be considered as well as its efficiency as an anti-corrosive agent. There are some substances used for protecting iron that act almost entirely as a covering; for instance, greasing iron is, when frequently renewed, effective in preventing further corrosion, because the humidity of the air cannot have contact with the surface of the metal. It has no particularly preservative chemical action with the metal, although, if it contains acids, it may promote corrosive influences, it being chiefly a mechanical protection. Paints should not be so thick that they cannot be easily spread, for continuity and evenness of the film are very important; or be so made that they have to be drawn out, which causes inequality of the coat; or be too liquid, for then they will be too thin to be an effective protection.

To sum up, the principal properties required in an anti-corrosive paint, or that spread on bare iron or steel, may be stated to be :—

(1) That it shall, without deterioration, have the property of suction, so as to extract and absorb any moisture or dampness that may be present upon the plate, or in the pores of the surface metal.

(2) That it is perfectly homogeneous, complete incorporation of the ingredients of which it is composed being attained.

(3) That it will form a hard, smooth, even, glossy, and uniform coating, impervious to air and moisture, thereby preventing corrosion, and have no injurious effect on the iron or steel, or the second coat of paint.

(4) That it adheres very tightly, evenly, and firmly to the bare iron or steel plate, and yet possesses the power of elasticity sufficiently to prevent cracking, scaling, or blistering when a structure is strained or vibrates, as bridges subject to a quickly rolling load, or ships, also from any contraction or expansion of the metal.

(5) That it presents such a surface that the second coat will easily spread upon, and very firmly adhere to it without affecting its quality or setting up any deleterious action, and without blistering either of the coats, or becoming cracked or chipped from vibration.

(6) That it dries reasonably quickly without any sacrifice of durability, in order that the second coat may be applied within an hour or so after the first coat. When this can be done, the number of painters has but to be increased, according to the area to be painted, to cover a structure in a few hours.

(7) That it can be easily applied, is ready mixed for use, only requiring to be thoroughly stirred, and that no thinning or heating is necessary; for, if heating is required, there is considerable difficulty in producing a fluidity sufficient for laying on the paint smoothly and evenly.

(8) That it shall not be deleteriously affected by climate or the usual variations of temperature.

(9) That its life, i.e., the time during which its qualities remain unimpaired, or only reasonably impaired from unavoidable exhaustion, should be considerable; and it should retain its preservative qualities until the established law of the decay of all materials has naturally become too apparent in it.

(10) That only two coats are required, and not more than three when it is exceptionally exposed; that it flows

U

easily, spreads readily, and has considerable covering properties.

(11) That its weight is not excessive, and its cost moderate.

(12) That it prevents galvanic action, and is an insulator of electricity.

(13) That it shall not be explosive, produce explosive vapour, or be readily inflammable.

The requirements cannot be placed in any order of importance, for all are necessary.

CHAPTER VIII.

NOTES ON PAINTS FOR METALLIC STRUCTURES.

THE preservation from corrosion of iron or steel in exposed engineering structures by a coating of some anti-corrosive substance is a matter of importance, see articles written by the author of this book, which appeared in *Iron*, March 31 and April 7, 1893, on the subject of 'Paints for Metallic Structures,' "for the durability of ironwork greatly depends upon the surface of the metal being properly cleaned and prepared before painting, the protecting substance being of the best quality, and the method of painting being properly conducted. Here it is advisable to first direct attention to the injurious and wasteful system occasionally practised of attempting to prevent corrosion and dispense with reasonable maintenance and care by increasing the number of coats of paint. No matter how many coats are put on, they cannot arrest incipient oxidation, or that on the surface of a plate, and when a recommendation is made to apply four or five coats of paint, perhaps this number may be necessary to preserve the *paint*, and make it somewhat difficult to see if its nature is destroyed, but certainly it ultimately fails to protect the metallic surface to any greater extent than two or three coats of anti-corrosive paint possessing good body and texture. The system of attempting to prevent corrosion by an increased number of coats, and therefore thickness of paint, is a deceitful one, for, by reason of the thickness of the paint, any corrosion on the surface of the metal may not show through the paint by discolouring it in spots or other-wise, as it will do if two coats are used. In the latter case

it can be remedied, being evident, but an increased thickness
of the paint may hide the oxidation that is proceeding, and
the mass of coats of paint may yet have sufficient adherence
to mask the decay until at length rust has formed to such an
extent as to become detached from the surface of the metal,
the paint ultimately falling off with the layer of rust attached
to it.   It is also a lazy practice, for it proceeds from a desire
to shirk the duty of reasonable inspection and maintenance,
and to give a false appearance of durability.   As a proof of
the truth of these statements, the maintenance of the Britannia
Bridge is effected by two or three men constantly on the work,
who attend to the slightest symptoms of local discoloration.
Thus oxidation can be at once arrested, and so long as this
practice is faithfully continued, it is impossible to determine
the end of the serviceable life of a structure so far as corrosion
is concerned ; whereas, if the surface be very thickly coated,
local oxidation would be invisible, although it was actively
proceeding, and decay would continue quite independently of
the mass of paint, which would, therefore, be the very means
of preventing the arrest of corrosion and preservation of
the structure that the extra thickness was supposed to effect.

It is well to discriminate between paints that are adapted
for painting for mere purposes of decoration, as; most house-
hold paints, and those that have to possess the far more
valuable and important properties of the preservation of
metals from oxidation and fouling, for they must not contain
anything injurious to the metal, and it is necessary that the
materials should be of the best quality.   It is preferable if
the paint proposed to be used be analysed to ascertain
whether there is anything in it deleterious to the metal to
be coated.   In the former case, any cheap substance will do
that takes up colour, presents a brilliantly glossy and smooth
surface, and is capable of being easily applied by unskilled
labour, so long as it makes things look pretty ; but, in the
latter case, a very hard problem is required to be solved, and
one that cannot be said to have been entirely conquered.   No
paint is the best to use under all circumstances, and those

manufacturers who have a real knowledge of the subject will be the first to allow that the nature of the paint must to some extent vary with the object for which it is to be employed, and the conditions under which it has to be applied. Thus, a ship paint that may succeed in the northern waters may be almost useless in tropical seas, and a paint for metals be unsuitable for a wooden or stone surface. There can be no such thing as universality in paints. If such property be claimed, it is simply an appeal to the credulity of the inexperienced, for it is advisable or necessary to know the nature of the surface to be covered, the especial object for which it is to be coated, whether the paint is for a submerged or an unsubmerged structure, and, if submerged, whether in salt, brackish, or fresh water ; if it will be exposed to strong acids, sewage, or the smoke or vapour of engines ; the degree of heat or cold to which it will be subject; whether the surface is clean and unpainted upon which it is to be spread ; and in what country and climate the work is situated, are some questions which, when answered, will cause variations in the preparation of the paint.

By way of illustrating the importance of a paint or protective coating being especially adapted to the surface to be covered, it may be mentioned that in the case of the cables of suspension bridges, consequent upon the variations in the length of the wires at different temperatures, and also in the curvature, it has been found that ordinary paint of various kinds, although it may be effective for other structures, is not so under such circumstances, because it does not tenaciously adhere to the wires under such trying conditions, but becomes fissured ; water or moisture is therefore admitted, and, moreover, it is confined, and thus corrosion is promoted instead of prevented. A solid hard coating is not to be desired for wire cables, or any work especially subject to vibration, but one that has much elasticity and adhesiveness to the metal, as in many other familiar examples which there is no occasion to particularise. Of all cheap and well-known substances used in paint, tar has been found to be the

best in such cases, as it does not crack or blister. It also
penetrates into the interstices between the wires forming a
cable for a suspension bridge or other purpose, and, as it
becomes sufficiently solid in cold weather, in addition to
being an efficient covering under ordinary climatic conditions, it protects the wires forming a cable from deteriorating
influences, notwithstanding variations of length or curvature.

Metallic paints and preparations of the carbonates and
oxides of lead or copper, can hardly be said to be entirely
anti-corrosive, because a pure oxide when applied to other
metals assists decomposition, and, further, as a metal is
electro-positive to its own oxide, electric action may be set
up to the particular detriment of the metal. It is question-
able whether boiled linseed oil of excellent quality, no oil
being used containing any sulphuric acid, when thickened
with pigments of earthy ochres, is not a sufficiently good
paint for ordinary purposes, such as railings, lamp-posts,
gates, and similar structures of inferior importance, but
when the preservation of surfaces exceptionally exposed to
deteriorating influences, such as corrosion or fouling, has
to be considered, other and more powerful protective coverings and compounds are required, not in the shape of a bulky
and merely mechanical mixture of different materials, but a
chemically-combined substance which will not separate, peel,
or allow moisture to penetrate; it also being free from
deleterious ingredients, such as copper, mercury, sulphurous
compounds, acids, or anything which produces galvanic
action, for it should have no chemical action on iron or steel.
Ordinary oil paints may be sufficiently preservative for iron
or steel which is under cover, but for bridges, roofs, landing
or promenade piers, and similarly exposed structures, some
more durable coating is required; for all oils, resins, and
gums, when exposed to general atmospheric influences and
to the sun's rays, oxidise, and ultimately become reduced to
dust, but long before this, the last stage of decay, they have
ceased to be an efficient protection to the metal.

Common white paint is frequently a mixture of white lead, oil, turpentine, and driers; litharge, sugar of lead, copperas and white copperas being chiefly used for this latter purpose, the copperas tending to indurate the paint. If linseed oil is employed, and it should be in preference, it is sometimes boiled with litharge to make it dry quickly, and is usually called boiled oil. When glossy paint is wanted, more oil is used. Turpentine spoils the gloss to a great extent, but helps to dry the paint. Good and durable oil paints can only be made by using the best and purest materials, and it is obvious that any such paint cannot be produced very cheaply. Not only should the oil be pure, but the white lead or zinc-white should be finely ground, of even quality and composition, and be thoroughly mixed; but there is a difficulty in obtaining a chemical combination of ordinary commercial white lead or zinc-white with oil, and thus producing a proper opacity. If this is not effected, uniformity in colour may not ensue, and the paint will then not equally adhere to the surface, but will peel and allow air and moisture to reach it, the result being corrosion of the metal. It is open to question whether ordinary white lead should be used for ironwork, for, being a carbonate of lead, it is a compound of carbonic acid, &c., and carbonic acid is an active corrosive agent. It is better, if lead paint must be used, to employ the best and purest finely ground red lead, which dries quicker, for the colour can be toned down as desired.

Linseed oil extracted from the seeds of the flax plant is the best oil to use when boiled, and is nearly the heaviest of the fixed oils that can be economically employed for paints, and is also one of the quick-drying oils. By boiling or heating, it acquires increased consistence, and it then dries somewhat more quickly. It possesses the property of adhesion in a marked degree, and has several other qualities which cause it to be the oil for oil paints, if they must be used on ironwork ; still, no oil paint can be said to be either durable or satisfactory for outdoor ironwork, for all oil paints become oxidised—although pure oil does not oxidise iron or steel—lose their tenacity, blister, flake, crack, and finally separate or weather away. The reason of the flaking, cracking, and separation may be said to be that the paint, being sufficiently oxidised, ceases to be an impervious coating, the metal then becomes oxidised, rust forms, and the paint is dilated till its elasticity is destroyed ; it consequently cracks, and finally flakes and falls off. It has been remarked that any substance used in paint which is an oxide may impart oxygen to the iron and promote oxidation, and, therefore, the paint, instead of being a protection, may be a more or less active source of supply of oxidising influences. This is a point to be remembered in using paints to cover outdoor metallic surfaces.

There are many different kinds of oils, animal ; vegetable, as linseed ; volatile, as manufactured turpentine ; with varying properties, some oxidising more readily on exposure to the atmosphere than others, and forming a thick gummy mass. When exposed to gaseous ammonia, such as that generated by manure heaps and decaying vegetable matter, the alkali combines with the oil, forming a saponaceous film, which, on being washed, or by the action of rain, dissolves, and so presents a fresh surface to the same operation, the oil thus being, as it were, extracted from the paint, which then cracks, flakes, and becomes fragile; and although, for other purposes, boiled linseed oil is the best to use in oil paints for ironwork, linseed oil, being the most easily saponified, is not to be preferred when ammonia is specially present. From the blistering, cracking, flaking, and falling away of oil paint when applied to a non-absorbent surface, as iron or steel, and the insolubility of oil in water, and, therefore, the absence of the property of suction, and confinement of any water on the surface of a metal, its adhesion to the metal is always likely to be deleteriously affected, and certain advantages when it is applied to wood are disadvantages in the case of iron or steel; hence any substance, not a promoter of corrosion, that possesses greater power of adhesion is to be preferred in paints for coating metals.

Paints made with euphorbium possess remarkably the power of adhesion to iron or steel. Solutions of asphalte or pitch also adhere very tenaciously, and they have sufficient elasticity not to crack on the expansion and contraction of the metallic surface; further, it has been claimed that they penetrate in a lesser or greater degree any oxide scale and envelop the particles, making them a portion of the paint, but this is a rather enthusiastic statement. Their solubility, it has been suggested, may be counteracted by mixing them with linseed oil. The following mixture has been well recommended: 2 parts of Brunswick black with 1 of red lead, finely ground, and thoroughly mixed with linseed oil. To this is added either natural mineral asphalt, pine pitch, artificial asphalt, gas-tar, or the residuum of petroleum distillation, if the crude oil has been distilled before any treatment with acid. It produces a very hard bright pitch, which is soluble in " once run " paraffin spirit, and makes an excellent, cheap, durable paint for ironwork in exposed situations. The tar paint used for the light-framed boarded roof which is covered with canvas, and protects externally the tubes of the Britannia Bridge, is thus made, the quicklime being necessary to neutralise the free acid in the tar. The composition is : coal-tar, 9 gallons ; slaked lime, 13 lbs. ; turpentine or naphtha, 2 or 3 quarts, the whole being dredged over with sand. The advantage derived from a carefully-spread coating of silicious sand dredged on any paint, while wet, is, that the particles of sand become embedded in the paint, and greatly help to exclude light and air, and prevent cracking, peeling, and blistering. A tar varnish often used is thus made : 10 gallons of coal-tar, fresh, so that the naphtha is present ; 2 lbs. of tallow ; 1 lb. of lamp-black ; $\frac{1}{2}$ lb. of resin ; 10 lbs. of finely-sifted fresh-slaked lime. The ingredients are thoroughly mixed, and the mixture applied hot, but not allowed to remain heated long before being used. A common tar paint is composed of 1 gallon of ordinary coal-tar, about 1 lb. of finely-sifted fresh-slaked lime, and about $\frac{1}{2}$ a pint of naphtha, to thin it sufficiently for easy application, as little as possible being used. It is heated and applied in the usual manner. When iron can be treated hot, in order to expand it and open the pores, paraffin raised to a proper temperature and poured upon it has been used with the view of pene-trating the metal to a sufficient depth to afford protection against oxidation, a suitable paint being subsequently applied.

As the application of tar and paints of that nature is likely to increase for iron and steel work in exposed or sub-merged structures not required to have an ornamental ap-pearance, and, as tar paint is by far the most important paint made on public works, and not bought ready for use from a manufacturer, it is here referred to at some length. If coal-tar be simply applied in the same way as paint, it dries very slowly, and may for a long time have a viscous surface, and therefore not be hard. The tar should be made hot, but not long before being applied; and if it requires to be thinned on cooling or otherwise, naphtha in about the proportions described can be used for that purpose. Petroleum is also occasionally employed, but the mixture should be applied hot to a metallic surface, and as soon as practicable after the metal is in its finished state, so as to prevent any oxidation, the surface being previously cleaned. It will then dry rapidly, and be as clean, hard, and harmless as any paint; but some finely-powdered fresh-slaked lime, sufficient to neutralise the free acid in the tar, should be incorporated with it, and, as the ordinary coal-tar of com-merce contains ammonia and phenic acid, it may be ad-visable in certain cases to have them removed by distillation and treatment with lime-water. It should not be spread upon a coat of oil or any other paint, as it will not become firmly attached to it, nor thoroughly dry and hard upon it.

The metal to which it is to be applied should be heated. The best degree of heat has not been established, but for cast iron, 300° F. has been considered sufficient, and from 600° to 700° F. for wrought iron, the latter being heated to a black red, the metal being dipped in the tar or pitch or asphalt mixture, which is to be kept at the boiling point, but the degree of heat should be carefully regulated so as not to affect the strength of the metal, especially in the case of steel, which it is, perhaps, well not to heat above 300° F., in order to prevent its ductility being seriously affected, or its becoming at all brittle. The heating of the metal is necessary : (1) to prevent incipient oxidation by removing moisture on the surface; (2) to expel the air from the skin of the metal, for, if this is not done, blisters will occur through its collection in one place, or from an accumulation of gases beneath the surface of the paint ; (3) to open the pores of the metal; (4) to attract the protecting coat; and (5) to obtain equal covering, adhesion, quick-drying, and hardening.

If the metallic surface is not heated, these properties are not attained, but there are often parts which are too large to be heated; they should be well cleaned and dried, and the tar paint spread on them when it is as hot as practicable. A recent specification for cast iron in bridgework required that the cast-iron pieces shall be cleaned and then heated to a temperature of 300° F., and immediately immersed in a bath consisting of 5 parts of boiling coal-tar, and 1 part of asphalt, and remain for not less than 30 minutes in the boiling tank, the temperature being maintained at not less than the boiling point during the whole operation. The result of such a process being carefully carried out is that the tar paint dries evenly and perfectly, has a smooth, varnished surface, and is exceedingly adhesive, becomes hard, and does not crack or become detached. Occasionally gas-tar is alone used ; if so, in submerged work, it will gradually wash off in a current or from wave action, for it does not combine and harden as properly made tar paint does.

If the metallic surface may be black, it is doubtful whether there is a better cheap coating for ironwork than tar-asphalt paint, freed from any trace of acid ; but if spirit be added, it is liable to make the paint too hard, and inclined to chip and become brittle. Coal-tar, naphtha, and resin are sometimes used, but successful application seems to show that a mixture of coal-tar and asphalt is to be preferred. Sir J. Brunlees specified that the rails for the San Paulo Railway should be coated when hot with a boiling preparation of tar and asphalt. On inspection in Brazil, some time after, they were found to be perfectly fresh, and he recommended, especially for foreign countries, that rails should receive a similar coating. The value of such a covering is now generally acknowledged, and, when properly done, appears to be universally approved, and it might be said that it is the only process that is so praised. It is, therefore, advisable, in order to prevent incipient oxidation, to have all iron or steel work that has to be shipped, if not required to be bare or ornamentally painted, especially if it is to be erected in tropical climates, cleaned and heated to a sufficient temperature, either by passing through a furnace or otherwise, and dipped in such a boiling preparation. If it is necessary to paint the structure in colours, tar paint cannot be used.

The metal having a temperature of about 200° F., or wrought iron that of a blue heat, can then be immersed in hot boiled linseed oil as soon as possible after it is rolled and sheared, so as to protect it from air and moisture; but this system has its disadvantages, as handling plates and pieces so dipped or brushed over with oil is difficult, and dirt is collected on the surfaces by the oily film, and therefore tends to prevent close-fitting built-up work, and tight riveting. A better method is to have the surface of wrought iron or steel carefully rubbed, and all blue or other scale so removed, or to dip the iron in a dilute acid bath, or to wash it with weak acid, the surface being subsequently properly neutralised by washing and drying, and the painting immediately done ; or the metal can be allowed to rust, and the scale be scraped off before painting, instead of by an acid bath and subsequent neutralising wash. The objection to oiling wrought iron does not apply to a casting, on which a coat of oil or paint should be spread directly it leaves the mould or foundry in order to prevent the commencement of oxidation.

Before shipment or leaving the works, a thin coat of anti-corrosive paint can be applied; but few paints will satisfactorily stand a tropical climate, with the exception of zinc paint and that prepared with the tropical issue euphorbium. For other than tropical climates, if the wrought iron or steel, before being erected, is cleaned with dilute acid, neutralised by a wash, scrubbed with iron scale, or otherwise cleaned, and steeped, when heated, in boiling linseed oil, and afterwards painted with two coats of red lead, toned, if desired, to chocolate or a similar shade, it may be considered to be sufficiently protected when the tar-asphalt coating cannot be applied because of its colour; but instead of the boiled linseed-oil bath and red-lead paint, a tropical gum paint, like that prepared from euphorbium, or one similarly endowed with the property of extracting and absorbing, without deterioration, any moisture from the surface of iron or steel, and having considerable elasticity combined with the necessary hardness and all the other anti-corrosive and antiseptic properties hereinbefore referred to as required in a paint for metals, is to be preferred for exposed and important constructional iron or steel work which is to be submerged or otherwise severely tried."

CHAPTER IX.

NOTES ON VARNISH, TAR, AND BITUMINOUS PAINTS.

WHAT are called varnish paints are now largely used, and are durable, glossy, and quick drying. They cover rather more surface than the same quantity of oil paint; a quick-drying clear varnish, instead of oil, being used with the compounds and pigments. It is sometimes made of gum juniper, turpentine, and pounded gum dammar, &c. They are variously composed, and contain hard gums, not poisonous or noxious to animal or vegetable life, although some may be distasteful. Only one gum, if it can be thus classed, has yet been used in paint, so far as can be ascertained, that naturally contains the necessary virulent poisonous principle, for it has to be introduced into most other anti-fouling varnish paints during the process of manufacture; copper, formerly largely used, but now condemned by the principal authorities, although employed in cheap paints, zinc, arsenic, mercury, &c., being introduced for this purpose. As these poisons are soluble in salt water, and more quickly so than the varnish paint, it follows the latter at last has no poison in it, or very little, hence the great value of a *poisonous* gum which retains its toxical properties until by the ordinary decree of nature its substance decays. Thus, as the paint wears away, a poisonous surface is still offered to life desirous of attaching itself, and its anti-corrosive and anti-fouling properties remain so long as the paint is on a structure, whether submerged or not, or on a ship's bottom; a great point gained.

An objection to anti-fouling varnish paints that have to be incorporated with a poison is that the mineral poisons, being heavier than the substance of the paint, sink to the bottom. This disadvantage does not exist in the case of a varnish paint which contains the necessary poison in the gum itself. Also, it is well to remember that in paint the spirits in it generally evaporate or dry up in about an hour or two, but the oils harden slowly, and may take days. Quantities of varnish are made by mixing linseed oil with a resinous substance, such as gum copal, gum dammar, and then diluting the mixture with turpentine, methylated spirit being also used as a dissolver and diluent. The reasons have been stated at the end of Chapter V. for doing varnishing in dry weather; but on exposed work, the polish or gloss gradually fades, to be followed by cracking or softening and general deterioration. The desired properties of varnish paints have been stated in the summing up of those required in anti-corrosive paints (see end of Chapter VII.). For the coarser kinds of agricultural implements, boilers, &c., and any surface that need not be other than black, a cheap varnish made from coal-tar is used; see the previous chapter. A very fine black varnish may be applied to any coarse iron surface by simply heating it to such a point as will cause it to decompose linseed oil, and then by brushing it over with this liquid. When it gets cold, the iron will be found to be covered with a fine, smooth, black varnish, which adheres very closely to the metal; however, such a varnish has not the antiseptic properties which tar varnishes possess.

A black varnish made on Dr. Lunge's method is declared to be excellent for iron. It is made by distilling gas-tar until nearly all the volatile products are expelled. The residual pitch is then dissolved in the heavier oils, or, if a very quick drying varnish is required, in light oils or naphtha. This varnish is the original tar, without the ammonia, water, carbolic acid, &c., which cause it to have a disagreeable odour, and to be so long before it is dry or hard. In a recent article on testing prepared tar, by Mr. C. Lunge, in the 'Journal für Gasbeleuchtung,' 1894, the fact is stated that prepared tar products are mixtures of coal-tar pitch with distillates of coal-tar in varying proportions. After the volatile products have been distilled from the tar, the hard pitch remaining in the still is difficult to remove, and its use, in the hard state, is very limited. It may, however, be utilised by adding to the still fluid pitch some of the heavy oils, which are of little value for other purposes. According to the quantity and quality of the oils so added, semi-hard or soft pitch, asphalt varnish for iron, or prepared tar, can be produced. In prepared tar used in roofing felt, &c., the volatile constituents having been removed, it is free from smell, and not liable to crack in drying. For both the manufacturer and user it is a great advantage for the benzol, phenol, naphthalene, and anthracene to be removed without decreasing the usefulness of the prepared tar, especially as by regulating the quantity of oil added, any required consistency can be obtained. Among the properties of tar of good quality are quickness in drying, covering capacity, resistance to weather, &c. Thick tar frequently contains air bubbles, which can only be separated by continued heating. Hard pitch often has a dull lustre, and does not soften in water at 208° F. The specific gravity was 1·2938 at 73·4° F.

Properly made and applied tar-asphaltum compositions, or tar, or a mixture of pitch and tar of such consistency that it does not run off a surface, or crack when dry, or bituminous paints, are anti-corrosive and generally preservative, and have been found to be especially adapted for iron-work in the tropics or hot countries; but black varnishes are used in which hardly any care is taken to expel or neutralise the acids or corrosive influences that may be in the tar, or to apply it carefully and in an approved way as has been described. For instance, common black varnish for iron frequently consists of gas-tar thinned with petroleum, used cold or hot just as may be convenient. Fine black varnish used for small machines, &c., is frequently made of fine lamp-black or ivory black in powder mixed with copal varnish. A varnish highly recommended to prevent rust in stoves is made of one pint of fat-oil varnish mixed with five parts of highly-rectified spirits of turpentine, rubbed on the iron or steel with a piece of sponge. It can be applied to bright stoves and even mathematical instruments.

Various mixtures and preparations of tar, pitch, tallow, resins, tar varnishes, and bituminous compounds, have been used, but those herein mentioned are reliable, although others may be so. As a rule, with paints of this kind, the simple mixtures are the most successful, for when many different materials are employed, considerable chemical knowledge is required to prevent the various substances acting deleteriously or even neutralising an individual good effect. As asphalt is so frequently alluded to as a preservative coating, it may be well to here refer to M. Malo's nomenclature in his excellent book, namely: (1) Asphalt is a natural product, a bituminous limestone, consisting of carbonate of lime and mineral bitumen, intimately combined by natural agency. (2) Asphalt mastic is the rock ground to powder and mixed with a certain proportion of bitumen similar to that originally contained in the rock. (3) Gritted asphalt mastic is asphalt mastic to which washed sand or river sand, free from all earthy matters, has been added. (4) Asphaltic or bituminous concrete is gritted asphalt mastic in a hot state, mixed with dry flint or other stone. (5) Bitumen is a mineral product found in asphalt rock in Trinidad and other places.

According to Boussingault, bitumen is composed of carbon 85 parts, hydrogen 12 parts, oxygen 3 parts. As stated by Mr. Delano,* it is therefore an oxygenated hydro-carburet. It is not gas tar, nor Stockholm tar, neither is it pitch from suets and fatty matters, or from shale or petroleum. Some asphalts are of a spongy or hygrometrical nature. As an analysis which merely gives so much bitumen and so much limestone may mislead, it is necessary to know the quality of the limestone and of the bitumen. It is well

---

* 'Minutes of Proceedings,' Inst. C.E., vol. lx.

to remember this in using asphalt compositions for anti-corrosive purposes. Asphalt is carbonate of lime naturally impregnated with bitumen in very variable proportions. Exposed to the atmosphere, asphalt gradually assumes a grey, almost a white tint, caused by the bitumen evaporating from the surface and leaving a film of limestone. No preparation of gas-tar is asphalt, nor is shale grease, pitch from suets, or Stockholm tar, bitumen. When asphalt is used, it should be ascertained that it comes from well-known mines, or the quality should be vouched for by an expert.

Pitch, being a substance of a most permanent character, and considered to be nearly free from chemical attributes, if, therefore, the pores of the metal are filled and the surface evenly and entirely covered by it, by many experts is considered the best preservative of iron or steel from corrosion. If properly applied hot to a clean unrusted surface, it will protect the metal for a long time provided it is evenly adherent and the continuity of the surface is preserved. For iron, steel, or wood, tar paints are more preservative than oil paints. Tar produced at gasworks at what would be called by gas engineers very low temperatures, consists largely of paraffins.

A series of experiments, at St. Elvoy, in Auvergne, since 1879, on the effect of preservative agents on mine timber,* showed the following results, the preservative value of oil paint being unity, and tar relatively higher in each case.

		Oak.	Pine.	Fir.	Beech.	Birch.	Poplar.
Tar	average	3·2	1·6	8·3	4·2	2·2	60
Oil Paint	,,	1	1	1	1	1	1

Three coats of the following oil colours were given. Pure white lead, white lead and yellow ochre, white lead and emerald green, red lead, white lead and lamp-black, white lead

* See 'Comptes Rendus Mensuels de la Société de l'Industrie Minerale,' November 1890.

and Prussian blue. The samples were examined every six months, one year and a half, two years and a half, three years and a half, and eight years and a half. The tar treatment was as follows:—Tar, mineral and vegetable, immersion for five minutes. Tests on fifty disks of each wood quarter inch in thickness. They were placed in a damp level underground, the prevailing temperature being about 75° F. The ages of the trees the timber was taken from were : oak, average thirty-one years ; fir, twenty-two years ; pine, twenty years ; birch, beech, and poplar, 5½ inches in diameter. This shows that oil and lead paints are unfit for warm, damp situations, and that tar paints are much more suitable in such places.

A solution of melted paraffin in three parts of heavy coal-tar oil, the mixture being kept fluid while being applied by immersion of the vessel containing it in hot water, is stated by Dr. T. Koller to be a most excellent wash or paint for · buildings very much exposed to atmospheric influences, and it may be useful, in order to prevent corrosive influences reaching the metal, when cast or wrought iron or steel are covered with a thin coating of brickwork or stone, as may the following processes. He also found by experiment that three parts of quicklime to one of common salt applied as a wash on brick walls, even when laid on with a syringe, stood remarkably well for four years, and was an efficient and durable preservative coating. He considered that this is due to the hygroscopic action of the salt, which absorbs water and permits of the speedy combination of the lime with the carbonic acid of the atmosphere. In the ' Imperial Institute Journal ' are recorded some recent experiments on the waterproofing of brick and sandstone with oils, communicated by Professor Liversidge to the Australian Association for the Advancement of Science, conducted with the view to ascertain the length of time brick and sandstone are rendered waterproof or protected by oil. They gave the following results. The oils used were linseed oil, boiled linseed, and the crude mineral oil known as blue oil, used for preserving timber.

The weatherings were made upon a flat portion of the laboratory roof, fairly exposed to the sun and weather. Good, sound, machine-made bricks were experimented on. The amount of oil and water taken up by the sandstone was very much less than that absorbed by the brick, altthough the area of the sandstone cubes was much greater thain that exposed by the bricks. Equal amounts of raw and boilled oils were absorbed ; the blue oil, however, was taken up im much greater quantity by both brick and sandstone, but by tthe end of twelve months the whole of the $13\frac{1}{2}$ ozs. of blue coil had apparently evaporated away and the brick had returrned to its original weight. The bricks treated with raiw and boiled oils remain unchanged. After the second oiliing, in November 1890, and exposure for nearly four years amd two months, they had practically retained all their oil, inaasmuch as they had not lost weight, and were also nearly imperervious to water. It was noticeable that the sandstone cubes itreated with raw and boiled oils returned to their original weights, but do not appear to have lost the beneficial effects of the oils, being also practically waterproof. Bricks hawe also been dipped in coal-tar to cause them to be less porouus, and, therefore, more solid and waterproof. They were simply placed for four hours in a large vat of hot, not boilinig, tar. By that time they were soaked full of tar, were blackc when taken out, and were allowed to cool. They weree then considered to be like rock, and when severely testeed in a blazing furnace at Detroit, they did not crack or exlhibit a flaw.

Some experiments, to prevent thick tar in the asccension pipes and hydraulic mains of gasworks, made by Dr. K̇unath, the manager of the Dantzic Gasworks, resulted in shiowing that in trying to dissolve the thick tar with variouus sub-stances, such as hot water, petroleum, linseed oiil, and turpentine, water and petroleum had no effect otherr than that of making the tar still harder. Linseed oil disssolved the tar slowly to some extent, but turpentine dissolveed it so quickly and completely that in a short time only thee small pieces of coal and dust remained at the bottom of the solution. It would be well to remember these results on applying tar on linseed oil bathed ironwork, or when turpentine is present in any paint used on tar, or tar spread on such paint.

Bitumastic enamel for ships' bottoms was used in sinking the cylinders of the Newburn Bridge, Northumberland, to stop the leakage of compressed air.   It was smeared over all joints and bolts, and was sufficiently tenacious to withstand air-pressure up to even 40 lbs. per square inch, and had elasticity enough to prevent cracking from vibration, and so had an anti-corrosive effect.   Asphalt mastic for the production of asphaltic concrete to be impervious, plastic, and to withstand the heat of the summer sun in California, mixed in the proportion of 25 per cent. of sand, 75 per cent. of gravel in 100 lbs., added to 10 lbs. of asphalt mastic, Mr. Purcell, B.A.,* found by experiment to be best.   The sand and gravel were put in at the upper end of a sand-drier, which was a horizontal cylinder 20 feet in length, revolving in a heated casing, and came out at the lower end heated to the required temperature, viz.: 310° F.   The material was then run into the mixer, and the asphalt added from a heating kettle at a temperature of between 280° and 300° F.   The experiments were made especially with the view to obtain an economical and effective impervious lining for reservoirs in earth, and earthen embankments, and may be also applied for protecting embedded ironwork subject to severe corrosive influences, and therefore reference is here made to it.   The asphalt mastic is thus made : To 100 lbs. of hard asphalt add 10 to 25 lbs. of liquid asphalt, specific gravity 10° Beaumé, free from water and volatile oils, below a temperature of 300° F. according to the hardness of the asphalt; the composition is then heated with moderate fires to 300° F., when it is used.   The asphaltic concrete is applied in the usual way with asphalts.   A heated roller is run over the concrete while it is warm, so as to smooth the rough

* Vide ' Minutes of Proceedings,' Inst. C.E., vol. cx.

surface and prepare it for the final coat, which is made and applied in the following manner. Three lbs. of liquid asphalt, capable of standing a 300° " flash " test, with specific gravity under 10° Beaumé, are added to 100 lbs. of hard asphalt; the whole is heated to 400° F., and the temper maintained by the addition of small quantities of liquid asphalt from time to time as the more volatile portion evaporates; to this is added between 10 per cent. and 20 per cent. of powdered carbonate of lime or sulphate of lime according to the peculiarities of the asphalt. It is thoroughly stirred whilst being used, and is poured over the surface of the concrete and smoothed with a salamander, a heated bar of iron, 2 inches in diameter and 8 inches in length, with an iron handle. One-eighth of an inch in thickness is the greatest thickness advisable for this coat, and it maintains its shape and answers all the purposes of closing the pores of the concrete, and rendering it impervious. A circular section of the material, 2 inches in thickness, was tested under a pressure of 200 lbs. per square inch, and proved absolutely impervious. It may be advisable to use such an impervious covering to exclude moisture and corrosive agents in some situations where iron is buried in the earth, as in piles, &c.

In the case of wire-rope suspension bridges it is found that the ordinary kinds of paint, i. e., those composed of white lead mixed with oil and a drying substance, while sufficiently good for iron used for the meanest purposes, are unfit for wire cables. The decay of several wire-rope suspension bridges in France is attributed to the use of these paints,* for when the paint dries it is found it does not afford a water-tight covering, consequent upon the contractions and expansions of the metal, vibration, and variations in the curvature of the cable. Fissures occur in the paint which admit water and help to retain it there. A fresh coat of paint put on such a delusive coating simply makes matters

---

* See ' Le Génie Civil,' 1881.

worse, as it hides for a time the corrosion that is proceeding, and neither remedies nor prevents the rusting of the cable. Experienced French engineers of high attainments have forbidden the use of such paints for cable suspension bridges, and also any rather quickly-drying paints. Coal-tar, the ammonia and phenic acid having been removed by distillation and treatment with lime water, has been adopted, as it becomes sufficiently liquid in hot weather to adapt itself to the expansion of the cables, and penetrates into them, and becomes solid in cold weather, closes the interstices, and protects them from damp. Paint prepared with euphorbium also possesses elasticity, tenacity, a similar power of remaining sufficiently moist, and is exceptionally anti-corrosive and anti-fouling in character. Drying-paints have been forbidden by some French engineers to be used for such purposes, and only those which solidify slowly and are even capable of becoming somewhat liquid again. In selecting a paint which retains some of its original moistness, it is important such moistness shall not be caused by a corrosive fluid, but be produced by a naturally anti-corrosive and anti-fouling liquid, such as tar free from acid, or the gummy juice obtained from euphorbia.

Some delicate experiments by Mr. A. P. Laurie on the durability and protecting powers of oils and varnishes, described in a paper read before the Society of Arts, London, showed that linseed oil, however pure or carefully prepared, cannot be relied upon to protect a surface from moisture. Resins seemed to better prevent its access, but they are brittle and perishable. Varnish was found to be of little use, and mastic proved to be best of the substances tried.

CHAPTER X.

NOTES ON ELECTRIC, AND SUPERHEATED STEAM COATINGS, THE
EXTERNAL CORROSION OF BOILERS, GALVANISING, AND
EXPLOSIVE PAINTS.

THE electric current has been used with the object of causing
iron and its compounds to resist oxidation. It is sometimes
called "bronzing" of iron, and it has been claimed that by
electrical process a few hours suffices to produce a most
excellent effect on articles of either steel or iron. It is referred to here more with the view of drawing attention to
the fact than for any other reason, as the results can yet
only be regarded as tentative, although it appears to be
probable they may be important; however, a long series of
experiments as to the effect of the process on the strength
and behaviour, &c., of the metal would be required before
the system could be considered as thoroughly reliable.

The chief engineer, Mr. Wehrenfennig, of the Austrian
North West Railway, made some experiments recently to
discover the original cause of injury done to locomotive
boiler plates by the formation of rust at temperatures under
and above 212° F. He found that a most complete protective
coating was formed on the plate when subjected to the action
of superheated steam above 930° F., known as Prof. Barff's
process for protecting iron from corrosion. About a temperature of 1100° to 1200° F., the action being continued for
some six or seven hours, has been adopted successfully.

It is agreed that many boiler explosions have been caused
by the external corrosion of boilers, which chiefly arises from
damp in any brickwork in which they may be set, therefore
it should be kept as dry as possible and be coated with a

waterproof composition. Moisture may be communicated to the brickwork, in addition to the usual means, by leakage through rivet-holes, longitudinal seams, and other defects, or want of sufficient care in preventing water or damp air reaching it, particularly if a boiler be wholly or partly below the ground level. Great width of bearing surface of a boiler on brickwork is liable to increase, if it be not the cause of, corrosion, for water will trickle towards it, and there remain. Boilers placed on a broad wall perpendicular to their vertical centre are especially liable to corrode, as then the lowest portion of their whole circumference is seated on brickwork. Consistent with sufficient bearing, the bedded surface should be as small as possible, in order that but little of the entire circumference of the shell is covered. Four or five inches bearing on fire-clay blocks so bedded as to raise the surface of the bed and prevent lodgment of water is recommended by experts in the examination of cases of external corrosion in boilers. Material of a hygroscopic character should not be used to join a boiler to the brickwork setting, and therefore fire-clay is considered far preferable to common lime mortar. The longitudinal seams should be so arranged that they can be inspected and caulked, and, therefore, should not be covered. It has been found, on inspection after explosions, that external and internal corrosion sometimes occurs on parallel lines. When there is any bend in the plate, grooving and channelling occurs, and then the internal corrosion becomes serious. In this book, the *internal* corrosion of boilers, which arises from so many different causes, cannot be profitably examined without entering into the whole question of the design, construction, setting, the character and temperature of the feed water, and the general management of boilers, which obviously cannot be done here. It has been estimated that the maximum *external* corrosion of a boiler is about $\frac{1}{16}$th of an inch in twelve months. By care, it may be much reduced and be almost imperceptible. The joints and any flaws in the boiler-plates are vulnerable places, and suffer most from heat and other causes.

Because of the differences between the temperature of the gases, about 240° F., leaving the economisers used in public hydraulic power supply, and that of the hot-well condensed water entering the economiser at from 65° to 70° F., Mr. Ellington has stated it has, in a few cases, been necessary " to renew the economiser tubes after three years' work, owing to the condensation on the outside of the tubes and their consequent corrosion."

With regard to galvanising, the only certain safeguard against corrosion of the wire in wire-rope suspension bridges, according to Mr. Roebling's specification for the East River Suspension Bridge, is stated to be "a coating of zinc," and as galvanised iron is so largely used, it may be well to mention that unless galvanising is effected on sheet or other iron of tough and ductile quality, its surface clean, even, and free from impurities—for dirt will prevent effective galvanisa-tion—the zinc, which must be pure, will not adhere properly to the iron, the consequence being that rust spots occur, and the metal becomes quickly corroded.

Why does galvanising iron protect it from corrosion under certain conditions? Because when iron is galvanised, i.e., coated with zinc, the zinc quickly oxidises, but its oxide being *in*soluble, it prevents moisture from reaching the iron underneath the zinc coating, nevertheless, zinc is affected by moisture, acids, and fumes, for sulphuric acid impairs and destroys galvanised iron by dissolving or deteriorating the zinc, therefore near gasworks, or any place where much coal is burned, as in smoky manufacturing districts, it will probably gradually degenerate, as also near or in the sea, as muriatic acid is present. It is also liable to abrasion, and when this occurs, any exposed parts of iron quickly oxidise. Experiments conducted by the French Government some years ago, went to show that water-tanks on board ship should be coated inside with tin, and not be of galvanised iron, as it was found the water, under certain various conditions, dissolved the zinc on the iron. It is said the process of galvanising by lead causes the coat to adhere

much more tenaciously than zinc does, and that the lead is not
affected by moisture, acids, and fumes, which so deteriorate
the zinc.

Lead coverings are used in manufactories to protect iron
vessels against the attacks of acids, and, as they usually
ultimately scale off, compound plates of copper and lead have
been tried with success, as it was found the copper and lead
become apparently intimately united and that in ordinary
working the strains and temperature changes produced equal
alterations of form.

Much thought and ingenuity have been directed to
forming a protective coating by chemical action on the
surface of iron, resulting in more or less success.   The
ferrous and ferric oxides are unstable, but there is another,
the magnetic oxide, which is stable, and when properly
formed and of sufficient thickness it is claimed prevents
further oxidation of the iron, and therefore preserves it by
one of its own oxides.   It is not, however, claimed that the
magnetic oxide will resist the action of strong acids, but
that it will prevent decay from ordinary corrosive influences,
and if a piece of the coating becomes detached the corrosion
will be confined to the bared spot only, and that it is a really
efficient preservative against corrosion.   The impediments
against its adoption for bridges, roofs, and similar iron or
steel work are that of expense, the difficulty of coating large
pieces of metal, and the unsuitability of the process for iron
that has to be bent, fitted, drilled or punched *after* treatment.
It is dissimilar to galvanising by means of dipping plates in
baths of dilute acid, thus removing all black scale, and by a
coating of zinc which, through its own rapid oxidation,
produces a surface that forms a stable in lieu of an unstable
oxide by another coating, or to the method of *lead* galvanising,
to which reference has previously been made.   Another well-
known process is enamelling iron, but it is principally
applicable to pots and pans, station names, and for advertising purposes, and need not be considered as useful for
engineering structures of importance, and unless the enamel

is fused on with the greatest care, corrosion sometimes occurs under the coating, and the enamel cracks and becomes detached. Here it may be well to state, Dr. Bloxam pointed out that if a piece of iron was first immersed in *undiluted* nitric acid for a length of time, and was afterwards dipped in *dilute* nitric acid, the latter had no effect on the iron, the stronger acid having apparently caused the diluted, and, therefore, weaker acid, to be innocuous.

A description of Gesner's method of protecting iron and steel against corrosion appeared recently in 'La Revue Scientifique.' It consists in forming upon the surface of the metal a double carbide of hydrogen and iron, which is very hard and adherent. A bar thus coated can be bent through an angle of 45° without disturbing the layer. In carrying out the process the articles are thoroughly cleaned from rust, but oil or grease need not be removed, although it is better to do so. A couple of gas retorts are placed alongside each other, and are raised to a temperature of from 600° to 700° C., corresponding to 1112° and 1292° F. The articles to be treated are then placed in these retorts for about twenty minutes, after which a current of hydrogen is passed through the retorts for forty-five minutes. A small quantity of naphtha is then introduced, the supply being maintained for ten minutes. It is then stopped, the current of hydrogen being kept up fifteen minutes longer, when it is stopped, and the retorts allowed to cool to 400° C., corresponding to 752° F., and when this temperature is reached, the doors can be opened and the finished product removed. The coating thus given has a bluish colour. It is not stated whether the coating is durable, or whether by tests the metals are proved not to have been deleteriously affected by the treatment. It would appear to be more suitable for small pieces than those of as large dimensions as bridge members.

Before proceeding to consider the subject of anti-fouling paints, it may be well to remember that there have been serious explosions on ships which have been declared to have been caused by certain of them; in fact, a few, as hereafter indicated, might be almost classed as "explosive" paints. However, with proper care, reasonable precautions, and without incurring any expense, there seems to be no reason why paints made of decidedly inflammable and volatile substances and liquids should not be used for coating ships' bottoms and submerged surfaces, and it is probably correct to say that a really efficient anti-fouling composition can hardly be made without using some substance or liquid possessing explosive power under certain conditions. Many things largely used daily, may be rendered explosive or inflammable, nevertheless they are used with impunity by many who do not even know that they possess those properties ; and in not a few cases the escape from imminent danger can be called nothing else than providential. According to the *Shipping World*, the death of Captain Rennie, marine superintendent, was attributed to the explosion of gas evolved from paint containing naphtha, benzoline, or shale spirit, being used in a confined space.

Explosions have been traced to have occurred from compositions containing naphtha, viz., that on H.M.S. *Hercules*, 1870, from the vapour of the mineral spirit accumulating and becoming mixed with a considerable volume of air in a confined space. Also on the *Coquimbo* in 1880. In 1881 H.M.S. *Dotterel* blew up by the explosion of the gunpowder magazine. It was considered to have arisen by the explosion of a mixture with air of the vapour of "xerotine siccative," a paint-dryer containing petroleum spirit. Explosions also occurred on H.M.S. *Cockatrice* and *Triumph* in 1881. In 1884, a cask of paint stowed in the forepeak of the *Hawarden Castle*, led to the occurrence of an explosion which killed one man and injured several others, the paint being subsequently shown to have contained 40 per cent. of coal-tar naphtha. Another occurred on the s.s. *Thorndike* in 1886, from the use of an anti-fouling composition also containing a large proportion of coal-tar naphtha. It is stated " these accidents arose from the incautious handling of a material which was not generally known to contain a liquid readily converted into vapour, and in that condition capable of forming a powerfully explosive mixture with air; and the same remark may doubtless be made in reference to accidents which have happened in the use of certain descriptions of varnish." *

Explosions have been traced to have been brought about by benzoline (petroleum spirit), mineral naphtha becoming mixed with air, the vapour so formed readily becoming ignited. Many volatile inflammable liquids form an explosive atmosphere, carbon disulphide, ether, and alcohol, for instance; benzoline and air, alcohol vapour and air, methylated spirit and air, are also specially liable to explode. As stated by Mr. Redwood, crude petroleum is a liquid freely giving off vapour at even a low temperature, for it contains the very volatile hydro-carbons, which, in the form of gasoline and benzoline, are separated in the process of fractional distillation to which it is subjected in the manufacture of the usual commercial products.

Petroleum vapour and air, and mineral naphtha and air have been proved to have caused serious explosions. Some mineral naphtha has been found to give off inflammable vapour freely at zero Fahrenheit. It is frequently necessary in the manufacture of paint to use methylated spirit as a solvent or diluent, but then any danger that may exist can be guarded against in the manufacture, and such paint containing any substance dissolved by methylated spirit, is only likely to explode or become inaudibly ignited by means intentionally used for the purpose. The practice of holding a naked light to a can which has contained such compounds is unnecessary and should be avoided, as also smoking and use of matches near any utensil containing such paint. Dr. Dupré, by experiments found that 1 volume of benzoline will render 16,000 volumes of air inflammable, and 5000 volumes strongly explosive, 1 to 3000 being scarcely explosive, but inflammable.

The vapour of oil is known to be very subtle, and oil is

* This information is derived from the valuable paper of Mr. Boverton Redwood, F.R.S.E., 'Minutes of Proceedings, Inst. C.E.,' vol. cxvi.

very penetrating, for it can leak through the plating of petroleum-carrying ships. When all *appears* to be quite safe, the atmosphere may be so caused to be in a very dangerous state, and oil tanks have been known to give off vapours, hours after being emptied, which, mixed with air, are dangerous. No petroleum which comes out of the earth is free from gas of an extremely explosive nature. Kerosene i.e., distilled petroleum deprived greatly of its gas and volatile oils, can be regarded in a different way to crude petroleum, which latter contains much " concentrated gas." There is a great difference between ordinary kerosene and heavy petroleum and the dangerous nature of inflammable liquids, such as crude petroleum, which contains considerable proportions of the more volatile hydro-carbons, and benzoline, but even the heavy oils evaporate gradually.

All bituminous paints are more or less inflammable, and there may be danger from them from evaporation and their vapour mixing with the air. The effects of volatilisation of any substance or liquids used in paint should be considered, for it may not only cause air to be readily inflammable or explosive, but the nature or good qualities of a paint may be destroyed. It is not advisable to use any substances or liquids of an explosive or highly inflammable nature in paints or compositions if it can be avoided.

CHAPTER XI.

ANTI-FOULING PAINTS, COMPOSITIONS, AND FLUIDS.

SOME of the principal characteristics of plants and animals which foul or encrust a ship's bottom having been examined, the means by which such encumbrances can be lessened or prevented have to be considered. Sir William White stated, in a recent paper read before the Institute of Naval Architects, that, " The evils of fouling in iron and steel ships are real enough. But, taking into account the great and increasing facilities for docking throughout the world, the improved anti-corrosive and anti-fouling compositions now available, and the moderate time and cost involved in cleaning and painting the bottoms of iron and steel ships, it becomes obvious that it is more economical, on the whole, to dock and clean as frequently as may be necessary to avoid serious fouling, rather than to incur the large additional cost of sheathing. Merchant ships, from the nature of their employment, are frequently in port, and all vessels of high speed are employed on fixed routes and regular services. Under these conditions, docking twice, thrice, or even four times in the year is not a serious matter, while with such frequent cleaning of the bottoms fouling cannot become serious."

Many compositions, paints, and fluids have been more or less successfully introduced with the object of preventing the fouling of ships' bottoms, and that of other submerged structures. The " food-substance " of all plants, and marine life which cause fouling has to be absent in any such paint, and, further, no life must be able to attach itself to the paint or anything upon which any fouling organisms can live.

Divers have noticed in many waters that fish will amuse themselves by eating grease from a ship's bottom. A successful anti-fouling composition must be highly destructive to all forms, minute or otherwise, of animal and vegetable life, and must not offer them food or a haven of rest denied them by the deep. It must also have a germicidal and toxical action, and retain such properties, or it will become valueless in time; however, its powers will decrease almost from the day it is applied, till its nature is destroyed and its anti-fouling quality is dissipated. Paints which possess and retain a smell distasteful to germinal and lower animal life have a decided anti-fouling character, because such creatures will seek some other resting-place. The dislike of such life for creosote has, among its other valuable properties, rendered it an excellent antiseptic for timber; however, it does not follow because a substance or fluid which is extremely destructive to vegetable life when present in considerable quantity, is so when it is in only a small percentage, for experiments have shown that certain liquids with carbolic acid will support fungi.

If a paint does not possess the power of paralysing the efforts of all marine life to adhere to a surface covered with it, it has to rely for its anti-fouling efficiency on its slowly becoming soluble in water, so that on life attaching itself it soon falls away. Any such property must of necessity be at the expense of its original thickness and durability. If an antiseptic or anti-fouler is used which quickly liquefies, consequent upon its avidity to absorb moisture, it is useless as an anti-fouling element in paint. Carbolic acids and some other antiseptics evaporate rapidly. What is wanted is an anti-fouling paint which *retains* as long as practicable its anti-fouling and antiseptic character, and is not volatile. In brief, the value of a paint depends, in great measure, upon its antiseptic and anti-fouling properties being fully retained in as non-volatile and durable a manner as possible.

Compositions may be divided into two classes, namely: those which have a soft surface so that anything adhering

Y

for a time ultimately becomes detached with a piece of the paint to which it was attached. The other, the varnish paints, are the more durable, as they become hard, do not dissolve, and rely upon the poisonous matter to prevent the attachment of all life, proceeding on the principle of prevention being better than cure, and these now hold the field among anti-fouling compositions. If a composition is of the nature of an hydraulic paint in the sense applied to lime and cement, it is generally a decided advantage, provided, on the surface setting and becoming hard, it retains its anti-fouling properties, and is sufficiently elastic and tenacious not to crack or become detached. A smooth hard surface is necessary or it will increase in roughness. It should not be soft or pulpy, for then friction is greater, and life can easily attach itself, or on contact will be retained. An anti-fouling composition must not offer a soft bed of decaying matter upon which animalcula can fatten and prove an attraction for higher orders and classes of marine animals. It is, therefore, necessary to know the habits of marine animals and aquatic growth, whether minute or not, that adhere to ships' bottoms, to successfully prevent their attachment, and some of their characteristics have been referred to in Chapters I., II., III.

A composition that will extract and absorb any moisture that may be in the surface metal, air and water seal the surface, adhere uniformly and firmly, present a smooth, hard but elastic, glossy and durable coat, be unaffected by the sun's rays, vapour, damp, moisture, or water, acids, salts, and anything that will deleteriously affect the metal, in addition to being of such a noxious character that no animal or vegetable life will adhere to it or touch it, is that which is required; and a really efficient anti-fouling and anti-corrosive paint must possess these properties and those enumerated at the end of Chapter VII., and subsequently in this.

A paint or composition that may preserve metal when in air may be useless or very ineffectual if submerged, especially in sea water. A parallel instance may be here referred to,

viz.: a successful process of preserving wood in air by immersing it in a weak solution of corrosive sublimate, i.e.
the bichloride of mercury, failed, or was only partly successful, when the treated timber was immersed in sea water.
The action of the chloride of sodium in the sea water, which
was not present in the air, was considered to have rendered
soluble the combined deposit in the wood. Mercurial salts
were therefore abandoned for preserving submerged wood.
It is well to remember this in selecting a composition.

Occasionally, by the exacting proprietor of a small pleasure boat, an anti-fouling composition is expected to prevent
any slime adhering to the surface. Until the laws of nature
are altered this cannot be effected, any more than dust on
land. Any one demanding such a requirement had better at
once commence improving (!) nature, for *after* dust has been
abolished, no doubt an anti-fouling composition will arise
to which nothing will adhere, *but not before.* At present,
chemistry has failed to prevent such sliming of the surface,
and mechanical means, such as washing and brushing, must
be employed to remove the slimy film, which should be
easily washed off or removed, and the painted surface then
be found to be fresh and even.

The chief requirements of a shipowner as regards an anti-
corrosive and anti-fouling composition, and its value in
influencing a ship's earning capacity may be said to be :—

(1) To prolong the serviceable life of a ship as long as
possible.

(2) That a ship should be capable of making quick
voyages.

(3) That it should only be necessary to occasionally dry-
dock, and not to have to scrape and repaint after each long
voyage.

(4) That the plates should be preserved by the paint from
corrosion and fouling of every description.

(5) That the paint should be cheap, ready for use when
stirred, no heating be required in order to produce fluidity,
or oils, liquids, dryers, &c., have to be mixed with the

ingredients; that it should be easily and inexpensively applied without highly skilled labour;] that it should dry quickly, and in doing so become hard, yet elastic.

(6) That the surface shall not blister, but be very adhesive, smooth, durable, even glossy and slippery, so as to reduce to a minimum the friction in passing through the water and prevent a reduction in the speed; and not become dissolved or wash off.

(7) That not more than two coats are necessary, and that they can be applied in one day, in order to save labour, the quantity of paint required, and to lessen the scraping when a ship has to be re-coated, thereby saving time and dock charges.

Choice has also to be made among the various compositions as to the best for the waters in which a ship will trade. A good ship paint should also have the following properties among others, viz.: the power of absorbing any moisture that may be on the surface of the iron or steel, and have the effect of extracting and absorbing dampness so as not to imprison it and cause corrosion under the paint; for if water is so confined, the application of a coating may cause the supposed remedy to be almost as bad as the disease, so far as the prevention of corrosion is concerned. The scraping and preparation of the surface is therefore of much importance, and a little extra care and expenditure will soon be repaid in a prolonged life of the ship and the paint covering its bottom.

It is seldom the plates are cleaned by burning and scraping, or acid and water, the surface being subsequently washed and dried; or by paint remover liquid, as can be done in the case of test plates. The usual operations with ships in dock is confined to scraping, scrubbing, brushing, cleaning by rubbing, or washing until the plates are exposed, and their surface is considered to be firm and clean, no attempt being made to extract any moisture that may be and generally is present upon the metal. Therefore a composition which forms an impervious coating, and also extracts and absorbs, without deteriorating any of its good qualities, any surface moisture, is to be preferred to one that does not possess that property. In cleaning the surface it is well to remember the deteriorating influences that have been in active operation during the voyage, in order to remedy their effects. What, then, are the chief results to be witnessed on an iron or steel vessel being bared in dry dock after long voyages in various waters, and when the virtue of the paint has become exhausted by time, or from its not having possessed sufficient powers of preservation to prevent corrosion and fouling? The conditions of ships' bottoms must vary very much. The following particulars and photographic examples will serve to classify them, commencing with a favourable and ending with an unfavourable condition. In each case the tests extended continuously during eight to nine months, March to December, April to January, and April to February. The colours of the paint when applied are mentioned in most instances, although colour has no effect in preventing fouling of any kind; but the general comparative appearance of the paint, apart from its condition, is of course affected.

1.

Condition of the Plate as it appeared on removal from the sea water—*Covered with a thin film of scum,* which washed off easily.*

Condition of the Paint on the fouling matter being removed—*Very good indeed.*

Condition of the Plate on removal of the paint—*Very good indeed.*

Appearance of the Painted Plate when raised from the water—

Colour:
Dark green.

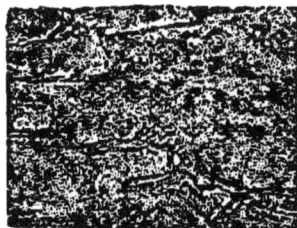

---

* The scum resembled pulverised coral, or slaked lime.

2.

Condition of the Plate as it appeared on removal from the sea water—*Covered with a thin coat of mud and scum.*

Condition of the Paint on the fouling matter being removed—*Very good.*

Condition of the Plate on removal of the paint—*Very good.*

Appearance of the Painted Plate when raised from the water—

Colour :
Bronze-green.

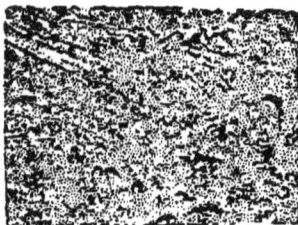

3.

Condition of the Plate as it appeared on removal from the sea water—*Covered with a thin coat of mud and scum, and a few barnacles.*

Condition of the Paint on the fouling matter being removed—*Some blisters, otherwise good.*

Condition of the Plate on removal of the paint—*Rusty where blistered, otherwise good.*

Appearance of the Painted Plate when raised from the water—

Colour :
Lead.

4.

Condition of the Plate as it appeared on removal from the sea water—*Covered with scum, mud, and a good deal of fine grass.*

Condition of the Paint on the fouling matter being removed—*Some blisters, otherwise fair.*

Condition of the Plate on removal of the paint—*Rusty where blistered.*

Appearance of the Painted Plate when raised from the water—

Colour:
Brown.

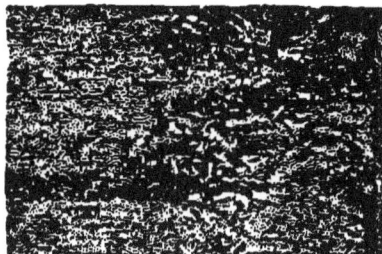

5.

Condition of the Plate as it appeared on removal from the sea water—*Covered with a thin coat of mud and scum, and considerably blistered, or the anti-fouling coat worn off.*

Condition of the Paint on the fouling matter being removed—*Anti-fouling coat worn off, and considerably blistered.*

Condition of the Plate on removal of the paint—*Rusty where blistered.*

Appearance of the Painted Plate when raised from the water—

Colour:
Lead.

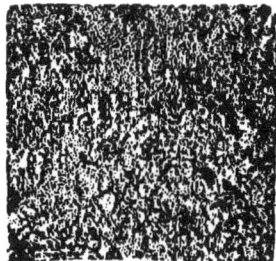

6.

Condition of the Plate as it appeared on removal from the sea water—*Covered with mud and scum and barnacles.*

Condition of the Paint on the fouling matter being removed— *A few small blisters, otherwise good.*

Condition of the Plate on removal of the paint—*Slight rust, where blistered, otherwise good.*

Appearance of the Painted Plate when raised from the water—

Colour:
Red.

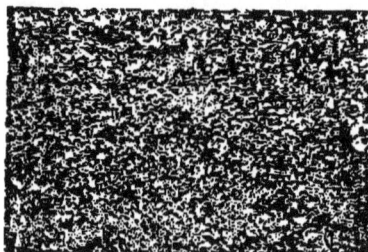

7.

Condition of the Plate as it appeared on removal from the sea water—*Covered with barnacles and oysters.*

Condition of the Paint on the fouling matter being removed— *Anti-fouling coat dissolved, paint washed off in cleaning.*

Condition of the Plate on removal of the paint—*Slightly rusty, otherwise good.*

Appearance of the Painted Plate when raised from the water—

Colour:
Chocolate.

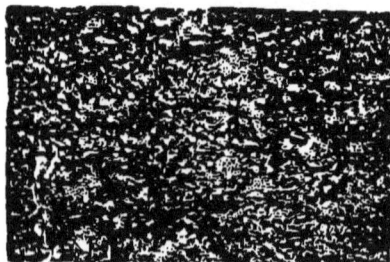

8.

Condition of the Plate as it appeared on removal from the sea water—*Covered with fine barnacles all over the surface.*

Condition of the Paint on the fouling matter being removed.— *Fairly good.*

Condition of the Plate on removal of the paint—*Only a little rusty.*

Appearance of the Painted Plate when raised from the water—

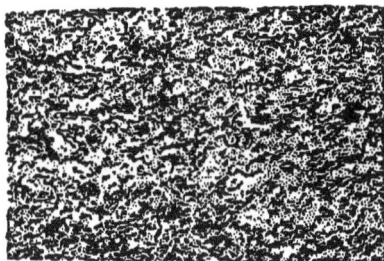

Colour:
Dark drab.

9.

Condition of the Plate as it appeared on removal from the sea water—*Covered with barnacles and oysters.*

Condition of the Paint on the fouling matter being removed— *The life of the anti-fouling and first coat of paint was gone.*

Condition of the Plate on removal of the paint—*Pitted badly, and covered with rust.*

Appearance of the Painted Plate when raised from the water—

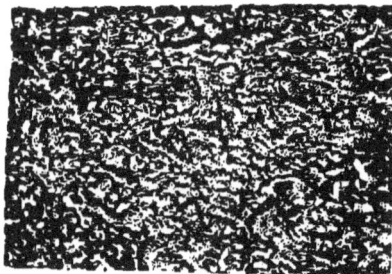

Colour:
Black.

10.

Condition of the Plate as it appeared on removal from the sea water—*Covered with mud, scum, grass; kelp had commenced to grow.*

Condition of the Paint on the fouling matter being removed—*Blistered very much, paint washed off in cleaning.*

Condition of the Plate on removal of the paint—*Badly pitted.*

Appearance of the Painted Plate when raised from the water—

Colour:
Crocus.

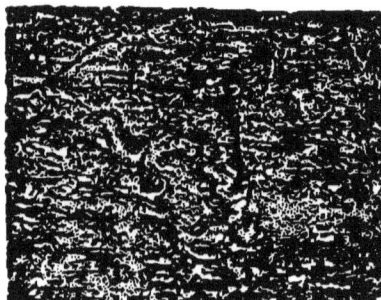

11.

Condition of the Plate as it appeared on removal from the sea water—*Covered with branch and shell coral.*

Condition of the Paint on the fouling matter being removed—*Hardly any paint on the plate.*

Condition of the Plate on removal of the paint—*Very badly pitted, and rusty.*

Appearance of the Painted Plate when raised from the water—

Colour:
Green.

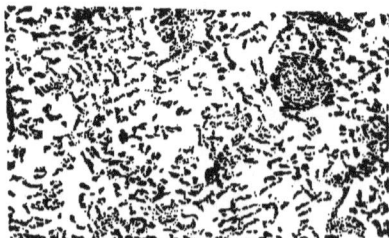

12.

Condition of the Plate as it appeared on removal from the sea water—*Covered with moss, a few barnacles.*

Condition of the Paint on the fouling matter being removed— *Very little paint remaining.*

Condition of the Plate on removal of the paint—*Rusty.*

Appearance of the Painted Plate when raised from the water—

Colour :
Crocus.

13.

Condition of the Plate as it appeared on removal from the sea water—*Covered with barnacles and moss.*

Condition of the Paint on the fouling matter being removed— *Very little paint remaining.*

Condition of the Plate on removal of the paint—*Rusty.*

Appearance of the Painted Plate when raised from the water—

Colour :
Red lead.

14.

Condition of the Plate as it appeared on removal from the sea water—*Covered with barnacles and very long moss.*

Condition of the Paint on the fouling matter being removed— *Paint nearly gone.*

Condition of the Plate on removal of the paint—*Slightly rusty.*

Appearance of the Painted Plate when raised from the water—·

Colour :
Brown.

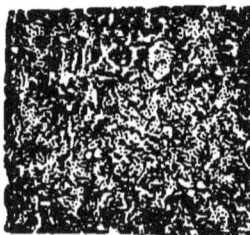

15.

Condition of the Plate as it appeared on removal from the sea water—*Very foul, covered with barnacles and moss.*

Condition of the Paint on the fouling matter being removed— *Paint gone.*

Condition of the Plate on removal of the paint—*Very rusty.*

Appearance of the Painted Plate when raised from the water—

Colour :
Black.

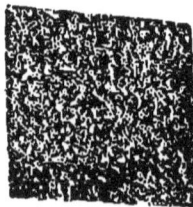

16.

Condition of the Plate as it appeared on removal from the sea water—*Covered with moss about 4 inches in length, and some barnacles.*

Condition of the Paint on the fouling matter being removed—*Paint gone.*

Condition of the Plate on removal of the paint—*Very rusty.*

Appearance of the Painted Plate when raised from the water—

Colour :
Bright Red.

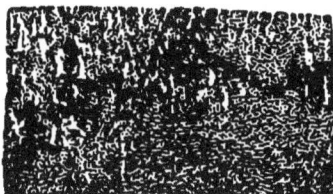

The results of some tests by the Navy Department, U.S.A., with some experimental paints were as follows, after 8 months 6 days' immersion at Portsmouth, N. H., U.S.A., Lat. 42° 59' N., Long. 71° W., of steel plates, 2 feet 6 inches by 3 feet, weighing 5 lbs. per foot, cleaned by burning, scraping and immersion in acid and water, then washed, and the plate dried.

A.

Composition of the Paint—*First coat, red lead; second coat, white zinc in oil, one side; the other side, white zinc in spirits of turpentine.*

Condition of Plate when removed from water after an immersion of eight months, six days—*Covered with thin scum; zinc and oil side clean, except as above. Very little kelp on side painted with zinc and spirits of turpentine.*

A—*continued.*

Appearance of the Painted Plate when raised from the water;, zinc and spirit side—

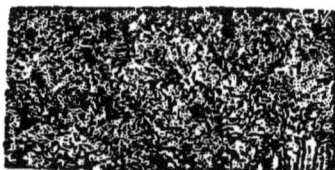

Condition of Paint on removal of fouling—*Zinc nearly all worn off on both sides.*

Condition of Plate on removal of paint—*Fine rust spots; redl lead not sufficient protection for iron.*

B.

Composition of the Paint—*First coat, red lead; second coat,', lampblack in spirits of turpentine.*

Appearance of the Painted Plate when raised from the water—

Condition of Plate when removed from water after an im-- mersion of eight months and six days — *Grass hadl commenced to grow; some blisters.*

Condition of Paint on removal of fouling—*Some blistered large? spots, bunches of rust.*

Condition of Plate on removal of paint—*Pitted and rusty.*

C.

Composition of the Paint—*First coat, red lead; second coat, yellow ochre in oil, one side.*

Appearance of the Painted Plate when raised from the water—

Condition of Plate when removed from water after an immersion of eight months and six days—*Badly rusted; grass had commenced to grow.*

Condition of Paint on removal of fouling—*Badly rusted.*

Condition of Plate on removal of paint—*Pitted and rusty.*

A perusal of the preceding and the following photographic examples of fouling and rust, &c., and the reports, will show to some extent what a really efficient anti-fouling and anti-corrosive composition or paint should be able to do, or prevent.

The following are some particulars of tests made with experimental paints on the U.S.S. *Speedwell.* Length of exposure, ten months and four days. Steamed about 338 miles on twenty-six different dates. For about six months remained fast to the wharf at the Navy Yard, Norfolk, Va.; 36° 45′ N., 75° W.

Composition of Paint—

    D. *First coat, $\frac{2}{3}$ red lead, $\frac{1}{3}$ zinc; second coat, plumbago, bees-*
        wax, tallow, and lampblack mixed with benzine.

    E. *Two coats of red lead.*

    F. *Two coats of white zinc.*

    G. *First coat, red lead; second coat, white zinc.*

Condition of Painted Plate—

    D. *Covered with barnacles; a little moss.*

    E. *A few barnacles.*

    F. *Very clean.*

    G. *Very little fouling; no barnacles.*

Condition of Paint on removal of fouling—

    D. *Last coat nearly all gone.*

    E. *Paint fair.*

    F. *Paint in excellent condition.*

    G. *Paint about half worn off.*

Condition of Iron on removal of paint—

    D. *No rust.*

    E. *No rust.*

    F. *No evidence of corrosion.*

    G. *No evidence of corrosion.*

Appearances of the Painted Surfaces on the ship being dry-docked—

    D   E   F   G

The principal properties required in an anti-corrosive paint or that spread upon the bare metal of a ship or submerged structure have been stated in Chapter VII. The principal requirements of the second or outer coat may be stated to be—

(1) That it shall be a certain preservative against fouling, whether proceeding from sea or bilge water, marine animals, such as crustacea, barnacles, corals, seaweeds, grasses, or aquatic growth of any kind.

(2) That it is perfectly homogeneous, complete incorporation of the ingredients of which it is composed being attained.

(3) That it will form a hard, uniform, smooth, even, glossy surface, thereby conducing to increase of speed, saving of time and consumption of coal, be impervious to air and moisture, and have no injurious effect on the first coat of paint, or the metal, should it come in contact with it.

(4) That it adheres most tenaciously, evenly, and firmly to the first coat without injuring it or causing it to blister, and possesses the power of elasticity sufficiently to prevent cracking, scaling or blistering on the ship straining or vibrating, and also from the surface being exposed to wind and water, or the sun's rays.

(5) That it reduces to a minimum the friction of the vessel in passing through the water by forming a very smooth, durable, hard, and glossy coat.

(6) That it dries rapidly without any sacrifice of efficiency or durability, in order that a ship may be quickly ready to proceed out of dry dock.

(7) That it presents such a surface that any coat will easily spread upon and very firmly adhere to it without affecting its quality, or setting up any deleterious action, and without blistering either of the coats, or becoming cracked or chipped from vibration.

(8) That it can be easily applied, is ready mixed for use, only requiring to be thoroughly stirred, and that no thinning or heating is necessary, for, if heating is required, there is

z

considerable difficulty in producing a fluidity sufficient for laying on the paint smoothly and evenly.

(9) That it shall not be deleteriously affected by climatte, or the usual variations of temperature.

(10) That its life, i.e., the time during which its qualities remain unimpaired, or only reasonably impaired from unavoiidable exhaustion, should be considerable; and it should retaiin its preservative qualities until the established law of decaay of all materials has naturally become too apparent in it.

(11) That only one coat is required, i.e., upon the first (or anti-corrosive coat previously spread, except when a vesssel trades in very foul waters, cannot be docked for an exceptionally long period, or other exhausting causes prevaiil, and that the paint flows easily, spreads readily, and hias considerable covering properties.

(12) That its weight is not excessive, and its coost moderate.

(13) That it prevents galvanic action, and is an insulatcor of electricity.

(14) That it shall not be explosive, produce explosivve vapour, or be readily inflammable.

It cannot be said that a *perfect* anti-fouling compoosition in every particular has been practically applied; but there are some that for a time very nearly fulfil this coondition. What are the principal agencies that destroy or aire distasteful to vegetable and animal life? (1) excessive colcd; (2) excessive heat; (3) electricity, provided it is sufficientlly powerful; (4) poisons; (5) explosions, sudden noise, amd loud reports, which will also frighten away fish and molluscca. It is hardly practicable to employ as agents Nos. 1, 2, 3, or 5, but poisons can be used, and they form the basis of anttifouling compositions. A less drastic remedy against tlhe attachment of marine life may be said to bo any substancce which causes paralysis, or is so distasteful or pungent thaat animal life will seek some other resting place. Anothter remedy is by utilising the secretions of the attaching animajls so that they form a non-adhesive soluble film on the coatinjg, with the result that the animal sets itself free almost as soon as it becomes attached.

Before proceeding to mention some substances which have been tried in ship paint, and more or less largely used, it may be well to consider the reason of the adoption of copper sheathing upon the bottoms of vessels, for it is sometimes thought it is to make a durable covering, so that repairs of a ship would almost be confined to those above the wind and water line. This is not the case, for it is employed to prevent fouling and the boring of marine worms, which in some waters, for instance, at Valparaiso, would quickly cause an unprotected wooden bottom to be honeycombed. It is somewhat singular that the copper sheathing should, by its own slow destruction, protect a vessel's bottom, and that marine animals, in trying to attach themselves to the sheathing, dissolve its surface very gradually, and themselves prevent the object they have in view, which is that of a resting and feeding place. The action may be thus described, for the copper does not poison or paralyse the efforts of marine animals to attach themselves to it, as paint made on the poisonous principle does, and in this respect it has no more effect than steel, iron, or zinc plate, for its value as a protecting shield results from a chemical change produced upon the surface of the copper by barnacles, molluscs, or worms, &c. The usual mode of attachment of mollusca, &c., to a submerged surface (*see* Chapter II.) is by means of a viscid or silky matter secreted or exuded by them. This secretion causes chloride of copper to be formed underneath that part of the animal which is temporarily attached to the copper plate, and in an immature state. As chloride of copper is a soluble salt, this area of attachment is no sooner so converted than the attaching matter dissolves sufficiently to cause the animal to be detached in consequence of its own exertions to attach itself. If this were the case with steel or iron, the plates would form their own protection against fouling by marine animals, but the salts which form from the latter metals are insoluble in water, and therefore the

z 2

animals can readily establish themselves on steel or iron plates.

The most virulent poisons, poisonous compounds, and fluids, such as arsenic, acids, &c., have been the chief agents introduced into compositions in order to prevent vitality of any kind ; but, by ceaseless effort, marine life at last finds a spot upon which it can live, and the erstwhile poisonous surface becomes innocuous. Other little inhabitants of the ocean cannot be employed to destroy life, or the presence of the gelatinous radiate animals, the tentacled and stinging Medusæ, whose combativeness and digestive powers are so marvellously great, might be desired.

No anti-fouling paint is indestructible, nor can it *fully* possess its properties for many months, a gradual and slow deterioration of the paint, which it is impossible to prevent, commencing almost from the time a ship is afloat, or a structure submerged. This leads to the question : How long will a composition retain its protective qualities ? The life of the paint is influenced, among other causes previously mentioned, by the foulness or purity, and the temperature of the water, and consequently by the location of the ports and seas in which a ship trades; to some extent the speed of the vessel ; whether almost constantly steaming, or in port for a considerable time ; and if a ship passes through drifts or currents having sand in suspension travelling at a considerable velocity, which have a scouring effect on the paint; or in a nearly stagnant sea, such as the Sargasso Sea, or great sea of weed, thus described by Humboldt : " Those evergreen masses of *Fucus natans*, one of the most widely distributed of the social sea plants ; driven gently to and fro by mild and warm breezes, are the habitation of a countless number of small marine animals "; or in comparatively cold waters, say N. of 50° N. The protective life of a good composition, properly applied upon an ordinary steamship, trading from, say, London to well-known ports in the Eastern Ocean, would probably be from six months to one year, and, in extreme cases, fifteen to eighteen months ; and in those N. of 50° N., should not be

less than about one year ; which means that a ship's bottom
would require to be scraped, cleaned, and fresh painted twice,
or, say, once a year—more often the former than the latter—
except when a vessel trades in northern waters. Many
steamships are, however, with great advantage, docked and
re-painted much more frequently. For vessels constantly in
latitudes N. of 50° N., the anti-fouling compositions absolutely necessary in more southern waters, and possessing
qualities enabling them, till exhausted by time, to repel
animal or vegetable growth of every description, are not
required in such strength ; although their anti-corrosive
qualities are equally necessary. Therefore a composition,
which may be most successful in northern and clean waters,
may by no means be effectual in more southern and genial
latitudes, say about S. of 50° N. It is also the same in the
case of submerged structures.

If a ship is docked frequently, it is no reason for using
a paint which is not decidedly anti-corrosive and anti-
fouling, for if it does not possess such properties, the plates
will become rusty, consequently to use any composition solely
because it is cheap and can be easily removed on re-docking,
is illogical, for protection from corrosion must be afforded,
and this, in the case of ships and submerged structures, cannot
be properly attained without due preservation from fouling ;
and it should never be forgotten that when corrosion has
commenced, it will proceed unless all rust is removed, and
the bare and solid metal coated with a really anti-corrosive
and anti-fouling paint that will maintain its protective
qualities in sea water. It is always much to be preferred if
a paint for a submerged structure, or a ship, is anti-corrosive
and anti-fouling, so that when from any cause the second
'coat becomes impaired or destroyed, the first is equally
anti-fouling, or sufficiently so to prevent fouling as well as
corrosion.

Respecting the numerous compounds which have been used
with the idea of producing an indestructible and impregnable
composition, whether for ships, submerged, or occasionally

wetted surfaces, it can be said that some of the most offensiwe
substances have been employed, and thick liquors possessimg
so strong an alcoholic odour as to force the remark from: a
painter that "it was a crying shame to waste good liquor iin
paint," the desire apparently being that so long as a compeo-
sition had an overpoweringly disagreeable or strong odouır,
it would cause marine animals to make tracks in anothıer
direction. The delicate sense of smell possessed by the lowrer
animals generally has been referred to in Chapter II. Theıre
is no doubt a composition, so long as it has an odour diis-
agreeable to the marine life which causes fouling, has an antti-
fouling effect, and therefore the property is not to be despiseed.

Among the substances which have been employed maıy
be mentioned, white and red lead, white zinc, spirits ıof
turpentine, turpentine, yellow and other ochres, challk,
gypsum, hematite, resins, oils of various kinds and qualitiees,
petroleum, petroleum residuum, tar products, creosote oılls,
bitumen, asphaltum, slaked lime, aloe decoction, wood· sorrcel,
oxalic solutions, saltpetre, naphtha and naphthalene, silicca,
manganese, colophony, lamp-black, gums of various kindls,
benzine, lac and shellac, euphorbium, graphite or plumbagzo,
verdigris, caoutchouc or indiarubber, asbestos, powderced
clay, grease, powdered mica, slate, alkalies such as sodla,
potash, and ammonia, lignum vitæ, ground coke, spermaceıti,
litharge, zinc, chloride of zinc, salts of tartar, shorts, soaıp-
stone, lime, tin, tungsten, plastic bronze, salt, arsenic, acicds,
laurel decoction, oxide or sulphate of copper, nitrate aınd
chloride of copper, sal-ammoniac, lard, sulphur, sulphate of
soda, Spanish white, lime-water, size, mercury, micca,
camphor, bismuth, alcohol, antimony, various vitriolic coım-
binations, paraffin, cellulose, hair, tin oxide, vulcanissed
caoutchouc, anthracite, beeswax, wheat flour, kerosene, tyıpe-
metal, saponified resin, heavy and light petroleum mixced,
flour of brimstone, animal fat, poppy-oil, cork, sulphate aınd
carbonate of lead, pulp impregnated with graphite, beeer,
steatite, tallow, glycerine, soluble gums, methylated spirrit,
yellow and brown soap, corrosive sublimate, acridine, whiich

iis said to possess intense pungency, acridity, antiseptic
]power, and freedom from loss by evaporation and the solvent
(action of water, peppers, cloves, and various other substances
(and liquids, in addition to all kinds of colouring matter.

There is no doubt many of these substances and liquids
(are of great value in the composition of a good anti-fouling
(and anti-corrosive paint or varnish, and that such composi-
(tions can hardly be made without using some of them, and
(that much thought and labour have been bestowed upon the
)manufacture. The substances in this by no means exhaus-
(tive list indicate that, except from some lucky discovery or
(ocular demonstration of its value, it is improbable that a
(really new and efficient anti-fouling and anti-corrosive
(ship paint will be introduced unless those possessing con-
(siderable knowledge of zoochemy and zoology, and the
(analytical chemist, the botanist, the mineralogist, and the
(engineer, act more or less in unison in order to comply with
the conditions that have been already enumerated respecting
the essentials of a good anti-fouling and anti-corrosive composition. Much disappointment and expense would be spared
if patentees of anti-fouling compositions would but study the
various practical requirements that have to be satisfied,
instead of becoming absorbed in *one*, which may have been
thoroughly mastered ; and if they would consider that some
knowledge of marine life is necessary in order to prevent its
attachment to a ship's bottom or a submerged structure.
This is written in no way to discourage inventors, but with
the view of indicating one of the pitfalls they may plunge
into in their natural eagerness to patent a discovery or new
idea. It has been the cause of not a few anti-corrosive and
anti-fouling paints having produced no adequate reward to
the inventor, and moreover has resulted in giving a hint to
others with greater practical knowledge of the subject and
unfettered freedom of ideas, who, by a calmer method of
digestion, were able to successfully supply that which was
necessary to make the invention a success.

Perhaps some despised poisonous plant or rambling shrub

of little or no apparent present value, may afford the durable toxical product or decoction that is required in an anti-fouling composition, for in the economy of nature everything has its use. At present, taking a broad view of the subject, it would seem that it is rather in the vegetable kingdom the magic substance is to be successfully sought. It must either poison, paralyze, or so deleteriously affect the delicate structure of marine animals as to prevent their attachment to a ship's bottom or submerged surface, or form a mixture soluble in water on the secretion of the marine animal being deposited upon it, as is the case with copper sheeting, and do this almost *without* deterioration. There are many familiar poisonous plants growing wild in this country, such as the deadly nightshade, the stinking hellebore, foxglove, monkshood, wolf's-bane, bryony, mezceerium, thorn-apple, &c. &c., which are either wholly poisonous, or partly so, whether in the roots, seeds, leaves, berries, or bark, &c.; and, in addition, there are the various poisonous fungi, without referring to other indigenous sources of poison. It is known that the Hottentots poisoned their arrows with the juice of the Cape blood-flower, and, looking to the success which has attended the application of euphorbium, perhaps the time is not far distant when some indigenous poisonous plant, or its issue, may be employed in the composition of an anti-fouling paint for ships' bottoms and submerged surfaces. To be entirely successful it must not only prevent the attachment of the higher orders of marine life, but also the algæ, the higher tribes of which embrace the seaweeds, and include the lowest of all the vegetative organisms, the protophytes or first plants; for in the marvellously wonderful economy of nature the higher orders of plants grow on mould formed by the decomposition of lower grades, hence the importance of a substance for anti-fouling purposes upon which the most minute life, of whatever kind, whether apparently mere slime or not, cannot even for a time exist, and then by its death become food for other and higher life.

Iron and steel ships have been sheathed with wood and copper, but coating them with anti-corrosive and anti-fouling composition is almost exclusively used, except for warships, on account of the expense. In the case of a warship, a very much longer period is likely to elapse between dry docking than with an ordinary vessel, and the much more permanently effective wood and copper sheathing system of preventing corrosion and fouling is therefore necessary in order that 'a ship may be able to keep the sea without serious fouling and loss of speed. Zinc sheathing, owing to *in*soluble salts being formed on it, consequent upon the action of sea water, has been found to produce roughness of the surface which promotes fouling. Copper sheathing experience has shown to be superior to zinc, and although more costly at first it is not so in the end. With regard to sheathing, direct communication between the copper and the iron or steel plating of a ship must be prevented, and any constant supply of sea water permeating between the copper skin through the wood and reaching the iron or steel plate and so setting up galvanic action between the copper and iron or steel; however, if the water which leaks through the sheathing to the plates remains, and is not subject to be removed by interminable fresh supplies, the corrosive action is very much reduced, being only due, as it were, to *one* attack, instead of illimitably repeated corroding onslaughts. · The wood sheathing should be free from acid and any decay, teak being generally used for this purpose, or the very object of the sheathing may be nullified and galvanic action be induced.

In a valuable paper on ' Wood and Copper Sheathing for Steel Ships, ' read before the Institution of Naval Architects by Sir Wm. White, K.C.B., LL.D., F.R.S., the Director of Naval Construction, the following paragraph occurs: " Experience has shown also that care is necessary even in the use on ordinary iron or steel bottoms *of anti-fouling compositions in which copper is largely used.*"

The constant poisoning of what may be called a film of water around a ship's bottom or submerged surface, is not

practicable as a means to prevent fouling, because a conn-
tinual supply of the poisonous element would be absolutely
essential.   There are many solutions which quickly kill the
lower forms of life, for instance, hydrogen peroxide, HOO_2,
has been found to be a violent poison for some forms ‹ of
animal protoplasm.   A solution of one of hydrogen peroxide
in 10,000 parts of water killed, in about a quarter of an hour,
all ciliate infusoria.   Anthrax spores which have resisted l a
four hours' exposure to a 0·1 per cent. solution of corrosive
sublimate, have succumbed in $4\frac{1}{2}$ hours to a 5 per cennt.
solution of chloride of lime.   The bacilli of cholera, typhoid
fever, and anthrax have been destroyed in one minute byy a
solution containing not less than 0·12 per cent. of calciuum
chloride.   Some such violent poisons as these, *without anny
corrosive or deleterious effect on iron or steel*, and retained j in
full power by the composition or fluid from the time a ship
is dry docked to its being painted again, is required ‹ to
destroy the infusoria that may attach themselves to a shipp's
bottom or submerged structure.

One of the most effective substances that have yet beeen
largely applied in the manufacture of anti-fouling and antti-
corrosive compositions is the strongly glutinous juice or gumm
issuing from certain plants belonging to the natural ordder
Euphorbiaceæ, specimens of which may be seen in Keew
Gardens.   Although gum euphorbium was formerly employeed
as a medicinal agent with correctives, its use is now almoost
obsolete except for horses, and it is only within about the
last twenty years that its remarkable preservative qualitiies
have been occasionally applied to engineering purposes.   Tl'he
genus Euphorbiaceæ is very numerous, there being somme
sixty-two species, six of which are natives of Great Britaiin.
It is a shrubby and herbaceous succulent, frequently armaed
with thorns, and has stalks from about 10 inches to 12 feeet
in height.   The juice is so acrid and irritating that it
corrodes and ulcerates the body wherever it is applied.   Tl'he
exudations of the species found at the Cape of Good Hoppe
have been used by the Hottentots for poisoning arrows annd

also ponds and springs, the branches of the euphorbia spurge being simply thrown into the water. Any animal partaking of water so impregnated quickly dies, but the flesh is not considered injured by the poison.

. The euphorbium of commerce, which is imported into Great Britain in large casks, is a gummy resinous substance which exudes from the stalks of the euphorbia plant. The drops are of irregular form somewhat resembling gum arabic. Upon breaking the lumps they will generally be found to contain little thorns, small twigs, flowers, &c., and some are hollow without anything in their cavity. The tears, as they are often called, are of a cloudy pale yellow colour externally, but are paler internally, and they easily break in the fingers, but are difficult to pulverise. The euphorbia plant grows in Africa, the East Indies, Mexico, Guatemala, Morocco, &c. The discovery of its probable utility in the manufacture of paint was made during a surveying expedition in Natal, in 1870, it being then noticed, when the clearing knives cut the euphorbia spurge, a strong glutinous juice or gum-issue adhered so firmly to the blades, that it was very difficult to remove, that iron so coated did not rust, and when immersed in sea water at Durban, no barnacles or marine life would touch it. In Natal, laths dipped in euphorbium juice were thrust into a white ants' (*Termes bellicosus*) nest mound, side by side with uncoated laths. In twenty-four hours the latter were found to be completely riddled, but the euphorbium-dipped laths were untouched. At St. Helena, it was noticed that even the moist traces of this insect on tin cases caused very speedy corrosion of the metal, showing its great destructive powers. Timber coated with euphorbium juice has also been severely tested against the ravages of the *Teredo navalis* with success. It is also found to resist considerable heat, the cold necessary in making artificial ice, and ammoniacal and chemical vapours, without blistering or scaling, or other injurious action. In the manufacture of paint, the euphorbium, after undergoing several special processes, becomes a

clear gummy juice of a darkish brown colour. This liquid is
mixed with the necessary colouring material, &c., to which it
gives a glossy appearance, while its own protective properties
remain unimpaired. Euphorbium paint has undoubtedlly
many naturally valuable qualities, especially when applied to
metallic surfaces, and the juice of euphorbium has a strong
affinity for iron and steel. It is a reliable preservative
against fouling, rust, and the corroding action of bilge water,
is easily applied without preparation, and has no injurious
effect on iron or wood or any substance generally used for
engineering purposes. It prevents marine worms, barnacles,
and vegetation adhering to a surface coated with it until its
protective qualities become exhausted by time; the bitter
pungent juice of euphorbium paralyzing all efforts of marine
and insect life. It maintains its quality in all climates,
and retains its virtue for a considerable time, and is most
tenacious, and yet elastic, for the prepared euphorbium can
be drawn out to a thin rod. It does not blister, crack, or
scale, and its viscosity is such that it will adhere to polished
steel, tin, zinc, lead, and any smooth surface, and also to
pipes buried in the earth. Brief reference to the use of
euphorbium has been made, because it is the most important and novel substance, naturally possessing the necessary
poisonous principle, recently introduced into the manufacture of anti-fouling and anti-corrosive compositions and paint
for submerged or peculiarly exposed structures; and it is
doubtful whether, for ships trading in any but very clean
waters, and for submerged structures generally, there is a
better material to employ, or one its equal for protecting
the bottoms of vessels or any metallic surfaces from fouling
and corrosion.

Asbestos has been used in ship paint, and is believed to
have been known to the Jews at the time of Moses, and
specimens have been found in vases unearthed from the
ruins of Pompeii. About the year 1780, it was proposed to
sheathe ships with it to guard them against the attacks of
marine boring worms, and towards the end of the last

century in England it was used, mixed with tar, as a paint for ships. Its composition varies, but it is a silicate of lime and magnesia, coloured by iron and manganese, and associated with water. It is insensible to the action of most acids, and to that of caustic alkalies. M. M. N. Melnikoff's monograph on Asbestos, in the Russian 'Mining Journal,' 1886, gives practical and very complete information for engineers on the subject.

It is probable cellulose, when protected from water or moisture, might be employed with advantage to close places especially liable to corrosion or fouling. In the French navy it is practically used, although not for this special purpose. Some is made from cocoa-nut husks, which is not pure cellulose, as analysis has shown it to have about 83 parts of pure cellulose, 10 of organic extracts, and 7 of ashes. It is considered the salts and the tannin secure the substance from decay and the attacks of insects. It is much lighter than cork, being about one-fourth of its weight. It is very compressible and elastic, and absorbs water rapidly by capillarity, and swells because of this absorption. It is compressed to half of its volume, and on a shot penetrating it its elasticity is such that an almost instantaneous obturation is effected. It is also used as a water-excluding briquette. The briquettes are enveloped in a waterproof covering to prevent absorption of water and increase of weight. Fourteen parts by weight of cellulose in a granulated state are mixed with one part of the fibre or filamentary part of the husk of the cocoa-nut. Cellulose may be said to be obtained by successive washings with cold or hot water, or weak alkaline or acid solutions, of a fragment of wood, until all soluble substances are removed ; a fibrous or cellular material remains, which is in effect the skeleton of the plant. The hull of the cocoa-nut contains a large proportion of cellulose in the form of a species of pith which holds together the fibres surrounding the nut.

A method of coppering ships' bottoms by electroplating has been practically introduced in the United States of

America.* The vessel is docked and her skin cleaned witth the help of a dilute acid which is applied by means of a water-tight bag of canvas and wire, large enough to enveloop the hull. After the surface of the metal is cleaned, the aciid is withdrawn, and its place taken by a solution of coppoer sulphate, electric currents being sent through the solutioon from a dynamo. In this process the hull of the ship formns the kathode, and the metal of the containing bag the anodde. The solution of copper-sulphate is thus decomposed, and a layer of pure copper is deposited on the iron. The shell (of copper can be of any required thickness, and it can be rce-newed. The process has been tried, it is said, with succeess in a vessel in New York harbour. The system of coatinig the surface of iron or steel by the electro-deposition of eithcer copper or zinc is, however, by experts not usually approveed for ships, the zinc making too rough a surface, and botth coatings being liable to be abraded and so rendered ineffecc-tual, for uniform protection from corrosion is absolutelly necessary, because, if local corrosion commences, it will sooon become severe, and extend in every direction, and the contacct of metals possessing different electro-chemical properties is generally undesirable.

Assuming it has been decided that a good compositioon has to be used on any vessel or structure, questions that wiill occur are—(1) how many coats should be applied, or be neceessary ? (2) should any part of a ship or structure have monre coats than another ? (3) if so, which, and how many ? Excerpt it is known a vessel will be in foul seas, i.e., those in whicch a ship's bottom or a fixed structure becomes quickly fouled, fcor a long period without docking to re-paint, two coats should bbe sufficient for ordinary purposes, as also for fixed submergeed structures, the first being anti-corrosive and the second antti-fouling ; the first coat generally being the thicker and heavierr ; but it is much to be preferred if *both coats possess anti-corrosiive and anti-fouling properties.* It is not easy to decide whetheer any part of a ship should have more coats than another, aas

* See ' Railway Review,' Chicago, 1892.

it to some extent depends upon the build, and the purpose for which a vessel was constructed; for instance, passenger packets, such as those used on the Dover and Calais and Ostend, Folkestone and Boulogne, Irish and Channel Islands services, have not the space between the load line and the light line that wall-sided, flat-floored cargo vessels have; but generally it is an advantage if a vessel has a third coat about the wind and water line, where there is the greatest external friction on the surface, or the light water mark and the load line, of say at least 4 or 5 feet in width, so as to ensure it is always well coated; and with regard to any other portion of a vessel, it is clear that there is no utility in having more paint at one place than another, unless it is subject to one or more of three causes, viz.—(1) special liability to foul; (2) for the plates to corrode; (3) for the life of the paint to be destroyed or the surface abraded. It is not probable that unanimity in answers to these points can be obtained, but with respect to *fouling*, perhaps vessels *foul* worst under the counters, and between wind and water, i. e., that part of a ship's side or bottom which is frequently brought above water by the rolling of the ship or fluctuation of the water's surface, and therefore is alternately wet and dry and exposed to the atmosphere: and with respect to *corrosion*, from the load water line to about the 5-feet water line, and least between the latter and the keel, i. e., upon the external or outside surface of the plates *corrosive* action not unfrequently increases from about the keel upwards, and whereas near the keel in a properly overlooked vessel it may be very little, it will probably increase from a little above the keel to the load-water line, the worst places being situated near the load-water, or wind and water, line to about 6 feet below it. The constant wash of bilge water is also most destructive, and the internal surface of the plates may be exposed to influences different to those of the outer surface, therefore no rule can be suggested for internal coating as the arrangement and build of a vessel will affect the magnitude of the corrosion. The outer surface, i. e., that in contact with the sea

water, being more constant in form; the extra band paintted
round a vessel for five feet or so near the wind and water liine
is an inexpensive provision against any extra fouling aind
corrosive tendency.

With regard to the quantity or weight of compositiion
required, two coats should be sufficient under ordinaıry
circumstances, for if a vessel's bottom is painted all over wiith
more than two coats it is doubtful whether it is advantageo)us
or compensates for the additional cost of the paint, labour,
and time in its application, and removal on re-paintimg.
It may also be found, when a third coat becomes worn out or
loses its virtue by time, that the second coat is in a decayiing
condition, and the first coat not in possession of its origimal
preservative properties, and that, from vibration or frictioon,
its adherence to the bare metal may have become defcctiive
from local causes, such as blistering, imperfect mixing aind
application of the paint, insufficient cleaning of the surface of
the plate, or from its ·becoming dissolved or porous.   T:he
best practice is to occasionally put on two fresh coats, in
preference to an extra coat or thicker covering and lcess
docking and inspection, the serviceable life of the plates aind
ship being thus prolonged.

The build of a vessel obviously has a considerable iin-
fluence on the quantity of paint required, so also has t!he
composition of the paint; but, of course, the value of a paiint
cannot be judged by its weight only.   The state of t!he
weather, the evenness and cleanness of the surface to `be
painted, the degree of rapidity of the painting, and the sk:ill
of the painter, also affect the quantity used.   The last thrree
are here considered as average and usual.   Respecting t!he
weather, when it is a cold and frosty or damp atmospheere
more paint is required than on a fine summer's day, and it
may be as much as 10 per cent., for not only does the paiint
not so easily adhere, but it requires more spreading,.and in
frosty weather to be almost dragged out instead of beiing
reasonably easy flowing.   With regard to the quantity of
paint required to cover one square yard of surface, t!he

governing conditions mentioned with regard to ships are equally applicable to ordinary painting. If the surface is curved, as corrugated iron, or much divided as in railings, gates, fencing, &c., or has numerous joints, more paint is used than for an even, continuous, flat surface ; the skill and carefulness of the painter, and the absence of undue rapidity of the painting, which causes waste from spilling and dripping, also affect the quantity employed. The weight required per square yard upon a metallic surface may range from as little, for the first coat, as $0 \cdot 11$ of a lb. to $0 \cdot 25$ lb., $0 \cdot 18$ to $0 \cdot 25$ lb. being most usual ; and 33 to 40 per cent. less per square yard for the second coat in the *heavier* paints is sometimes used. One lb. may cover 8 square yards the first coat, which is an extreme case. The author has known 2 lbs. to cover 15 square yards, one coat, with expert painters applying it on a large flat surface, and the same kind of surface to require 1 lb. of the same paint for every 6 square yards, or 2 lbs. for every 12 square yards, one coat. From 4 to nearly 6 square yards per lb. for one coat is the average. In ship painting, consequent upon the necessary rapidity of painting, rougher manner of application, dropping and spilling, about half a pound per square yard of the surface painted is required for *two* coats, and generally somewhat less, being anything from about $0 \cdot 45$ lb. to $\frac{1}{2}$ lb. for *two* coats according to the consistency and nature of the paint and the other influencing circumstances mentioned. It should be remembered that great covering properties of a paint may detract from its protecting and preservative qualities, and that there is a limit beyond which it is inexpedient to proceed. However, consequent upon the different ingredients used in paint and their varying consistency, no arbitrary quantity per square yard can be established, but something more than a mere wash or solution is required in paint for anti-fouling purposes, and good body and consistency are absolutely necessary, and equally so in an anti-corrosive paint.

For vessels with a rough surface, or large girthed, or with

a floor nearly flat, from 10 to 15 per cent. additionnal quantity is required to that for ships of ordinary build, annd usually when a vessel is not thoroughly scraped and a different paint is spread on that with which it was previouskly coated, from 5 to 10 per cent. more; but the better plan is † to have the ship's bottom scraped, as some compositions do naot adhere well to others, and may also act deleteriously one con the other and on the metal; in fact, such an application ι is offering an invitation to failure of the paint and deterioratiàon of the metal. In the case of some anti-fouling compositioɔns the quantities required for the first and second coat aare identical, but they range from the same to 50 per cent. morre, the first coat being 50 per cent. more than the second coaàt; thus if 2·10 cwts. were required for the first coat, 1·40 cwvt. would be the weight for the second coat, and so for any othɔer proportion.

As the weight of paint required varies, and especiallly according to the composition of the paint, a table of quaɯn-tities required might be misleading, although for one kiɾnd of paint it can be reliably constructed on the principle ‹ of tonnage to length of vessel. A minimum and maximuɯm quantity could be given, but it would be no absolute guiàde to the relative cost of coating a ship, for a light compositiàon would then appear to be the cheaper, whereas, taking inɯto consideration the relative value, quality, and body of tɛhe compositions, it might be the dearer. A simple weight teest will, therefore, generally be delusive. The least quantiàty for two coats for a vessel 170 feet in length and of 7C00 tonnage, painted 10 feet up, would be about 3¼ cwts., if 16 feet up, 5 cwts., the maximum about 3½ and 5½ cwtts. respectively. For a vessel 400 feet in length and 45G00 tonnage, painted with two coats 14 feet up, the minimuɯm quantity would be 14½ cwts., the maximum about 1G cwtts., if coated 21 feet up, the minimum required would be 17 cwtts. the maximum, 19 cwts. The approximate weight of a galloɔu of ship paint is 14 lbs., but, of course, it varies according ɫ to its composition.

At present, there does not appear to be any specially advantageous colour for anti-fouling paint. Although Professor Marshall Ward's, F.R.S., experiments on blue and violet light as germ killers show that all colours have not an identical bactericidal action; so far as the colour of ship paint is concerned, it would appear from the results of fouling of paint of various colours by marine life, that it has little, if any, effect. Makers occasionally recommend certain colours to be preferably used, and there may be reasons for this other than those of cheapness in manufacture, for some colouring matter may more readily become incorporated with the anti-fouling constituents forming the composition, and, therefore, the statement that such colours as salmon, purple-brown, stone, lead, dark-red, green, or black, are those which will be found to possess the most resistance to fouling and atmospheric deterioration is not to be disregarded. Unless for some special reason, there seems to be no advantage in ordering paint of any particular colour for a ship's bottom except it is to procure the cheapest, and one equally or more effective than a more expensive tint, or for distinguishing purposes. Consequent upon the colour of the anti-fouling ingredients in such paint, it may be impossible to make a pure white tint; a white with a yellowish tinge, or a light stone or grey colour being the lightest hue to be manufactured without some loss of the anti-fouling properties.

Having described the chief characteristics of the fouling and protection by paint of ships' bottoms, it is only necessary to briefly refer to other floating structures. With regard to lightships, buoys, pontoon landing stages, floating bridges, and similar moored structures, the chief points of difference between them and ships are, (1) it is known they will remain in the *same* waters, and it may be presumed under the same conditions; (2) fouling, *provided it does not assist corrosion*, is of little moment as compared with an unclean condition of a ship's bottom, for speed in vessels is of the greatest importance, and should not be reduced. It might even be said

2 A 2

if the submerged area of a floating structure was foul, i it would be conducive to the stability of a moored or anchorered vessel, the magnitude of the fouling not being likely to b be such as to seriously affect the displacement. As a shiip should always be coated with a composition capable c of preventing fouling in the *worst* waters in which she tradeies, attention must be directed to the anti-fouling coat as well a as the anti-corrosive coat, whereas in moored structures, thhe anti-corrosive property of the paint is the *chief* consideratioion.

The parts of a vessel which are likely to become foul thhe quickest have been indicated, and in an iron pile structurire, such as a pier, the most fouling generally occurs at abowut low-water mark, principally because at about that depth thhe water is usually not so disturbed ; marine life is not left on a a dry surface, and yet it receives fresh supplies of food at eacich tide, and, in addition, any matter near the surface. In ordder to lessen the fouling and corrosion of light pile structurees, such as landing or promenade piers, it is well if all bracinpgs are sufficiently rigid and strong as to act as either struts c or ties, and to reduce their *number to a minimum and increaiase the sectional area of each*, and also that all *external* flanges l be avoided, joints as described in Chapter XII. in the First Pa'art of this volume,* or a slot and stud arrangement be adopteced, or the external flanges made as few as possible, especialllly below low water, or where they cannot be frequentltly inspected to see if they become foul or corroded. If tlthe difference in the height of high and low water be moderatute, it may not be necessary to have any struts or ties betweeeen low and high water, and so lessen fouling, the groups ı of piles being braced at about low water, and at their tops ı in promenade and landing piers, the obstruction offered to tlthe waves being thus reduced, the thickness of the parırts increased to ensure that the bars, rods, &c., really act ı as bracings, and do not become a network of loose and distortcted struts and hanging ties consequent upon continual vibraticion

* 'The Safety and Preservation of Iron and Steel Structures.'

and their small area, and thus fail to afford that measure of
support to a structure which is supposed to be their joint
and several office, and prove to be a kind of grating to catch
any fouling matter that may be in the water. The flanges
and joints should be caulked or set in a durable cement, and
all external ribs, bolts, cotters, and turnbuckles be avoided
as much as possible; in fact, bracing that is submerged,
occasionally submerged, or exposed to wave action, should be
designed differently to that of bracing for structures above
water, such as bridges or roofs, or wind bracing, as it is
especially liable to deterioration from concussion and vibration caused by the waves and floating substances and
looseness of the joints, therefore the difficulty of keeping it
sufficiently tight to be effective and reliable is increased, and
corrosion and fouling at the most vital places are accelerated
by mechanical action.

The fatal disaster at the Morecambe Bay marine pier, on
September 9th, 1895, owing to the collapse of the landing
stage at the end, confirms that which had been previously
written in the First Part of this volume,* as to the serious
character of corrosion generally, and especially in metallic
structures of light construction, and the importance of bolts
being on no account used to sustain a girder, platform,
landing place, stair, lift, gratings, or anything upon which a
living load may rest, or the structure solely depend for its
strength or stability, and the desirability of altogether
avoiding the use of external bolts. The Board of Trade
report indicates that "the corrosion of the bolts was the
direct cause of the accident." The pier was constructed in
1868, and an extension made in 1872. In 1895, when the
collapse of the landing stage occurred, the words used were,
" The bolts were probably an inch in diameter when new,
but they were now practically eaten away, and had collapsed
beneath the pressure."†

Even when firmly bedded upon girders, and not supported

* 'The Safety and Preservation of Iron and Steel Structures.'
† See Col. Addison's, R.E., report to the Board of Trade.

by bolts, cast-iron platforms on piers should not be used for landing places or to sustain a load directly or indirectly, for in addition to their general unreliability, they may be broken by blows or shocks, and the metal may be changing in character from corrosion and decomposition, although apparently sound.

The fouling and corrosion of ships' bottoms and its prevention and reduction have been considered at some length, and that which has been written is equally applicable to lightships, buoys, landing stages and pontoons, pile piers, lighthouses constructed upon piles, dock gates, caissons, and similar structures, whether made of steel or iron, and therefore need not be repeated ; but it may be unadvisable to adopt iron or steel in dock gates when they will be subject to fresh and salt water, sewage and water, or the effluent water of chemical works, and to remember that caissons can be externally painted all over by reversing the caisson in the stops ; also that ordinary timber will not resist the attack : of marine worms, if it should be thought desirable to use timber instead of iron or steel.

CHAPTER XII.

NOTES ON PAINT SPECIFICATIONS.

WITH regard to specifications and the prevention of corrosion and fouling, it is not customary to particularise the properties required in anti-corrosive, or anti-corrosive and anti-fouling paint. The usual name of the paint or composition may be given, and is frequently nearly all that is written respecting it. Sometimes the sentence used is "paint of approved description and colour." This allows considerable freedom of action, unless by the word approved it is clearly stated it means the engineer's approval, and that his decision is final, for the very cheapest paints that can be made and are sold might be said to be approved, for they are largely used. It would be a somewhat vexatious task to declare a paint not approved, except for colour.

Almost all paints may be said to be anti-corrosive, for they cover a surface, but their anti-corrosibility may be only ephemeral, and in a very short time they may be found to be, if not promoting corrosion, certainly confining it. In Chapter VII. it has been stated that on examination of a large number of iron and steel bridges coated with iron oxide paint, red lead and white lead paint, i.e., ordinary lead and oil paint, rust was found *beneath the paint.* Such paints would generally be called "approved paint," and yet not prevent corrosion proceeding *beneath* a surface coated with it, although such corrosive action may be primarily caused by moisture in or upon the metallic surface being present at the time of spreading the paint, and so being confined by it, and not be initiated by the paint, but only assisted by it; for it

has been noticed in cases in which the metal had been oiledi
in the shops there was much less rust under the paint thann
where this precaution had been omitted. This points to thee
corrosion beneath the paint being principally caused byy
confinement of moisture aided by imperfect paint, and showss
the necessity of specifying that the surface should be properlyy
prepared, see Chapter V., before paint is applied, and thatt
the paint shall, without deterioration, have the property off
suction, so as to extract and absorb any moisture or dampnesss
that may be present upon the plate or in the pores of thee
surface metal. In Holland, especially, much care is takenn
to properly clean and prepare a surface before painting, andd
it would almost seem as if one of the national characteristicss
of the Dutch, viz., cleanliness, had not been forgotten by thee
engineers of that country.

It might be said it is somewhat singular that after thee
most exhaustive requirements as regards all other matters,;,
the preservation of the metal of a structure, even when alll
its parts are not accessible for inspection and re-painting, iss
in the specification sometimes dismissed in a few lines, suchh
as " The iron is to be painted on all surfaces before leavingg
the foundry, with two coats of metallic paint and oil, andd
with a third coat of lead and oil after the bridge is erected.'"
Such requirements cannot be compared with those observed inn
the Forth Bridge, where every practicable precaution has beenn
and is taken against corrosion and deterioration of all partss
of that colossal monument of engineering science; or withh
the constant care bestowed on that masterpiece of the day;,
the Britannia Bridge, which has been referred to in the Firstt
Part of this volume.*

The stereotyped phrase of "approved paint," or lead paintt
of good quality, which means such as any painter uses forr
the most mean and common work, is sometimes consideredd
sufficiently good ; and it may be the selection of the paint orr
the entire direction of the painting is assigned to some onee
whose knowledge of the requirements or even the object of aa

* ' The Safety and Preservation of Iron and Steel Structures.'

preservative coat on a metallic structure may be decidedly microscopical. The average foreman will declare for lead paints whose composition, as has been shown, is hardly conducive to anti-corrosive action, but it is " oil and lead paints " with many, or nothing. The reason is many know nothing of any other paint than oil and lead paints, which they have been accustomed to use, perhaps from their boyhood. If any corrosion on the surface of the metal, cracking, blistering, or peeling should occur it may be roundly said it was to be expected, for another coat or two should have been applied, which is sure to cause a consensus of approval as more labour and materials are required. Any really carefully-made paint of proved anti-corrosive and preservative ingredients is disliked, because it can only be made by experienced men, special processes, and expert advice; and to not a few it is almost rank treason to use any other than oil and lead paints. Any cheap stuff that will show some body, and take up colour, is made or approved, when it is impossible to buy the best materials of their respective kinds and properly manufacture a really durable anti-corrosive paint for the price at which it is required to be sold and delivered.

The system of advertising for paint and accepting the lowest tender on a specification merely requiring " metallic paint," " oil and lead paint," " white lead and oil paint," " oxide of iron paint and oil," " red oxide of iron in oil," " thick red-lead paint," " two coats of oil paint, the first coat to be red lead and oil "; without any or with slight knowledge of the character and quality of the ingredients, has a vagueness which is most refreshing to a dealer, but can hardly be termed a satisfactory method of procedure, or conducive to the durability of a metallic structure.

Some specifications contain the words " the iron to receive two coats of boiled oil and red lead," " two coats of oxalic paint," " thick red-lead paint," " good red-lead paint," " best metallic oxide paint," " approved anti-corrosive paint," " paint of approved description," best linseed oil and lead paints," " the paint to be of description and colour approved by the engineer."

More carefully-drafted specifications contain words t,to the effect that the paint must be "anti-corrosive paint app-proved by the engineer," "the paint to be made of the bes,st materials and liquids of the respective kinds and to bbe approved by the engineer or his inspector."

In some specifications requirements are made as t,to cleaning the surface as well as to the paint. They m ay bbe worded to the effect that "all the ironwork shall, as soon a as possible after being cleaned, be painted with one coat o of oxide of iron paint and oil approved by the engineenr. Immediately after the erection of the bridge, the ironworlrk shall be thoroughly cleaned and be painted with two coat,ts of white lead and oil, of such quality and tinted as thhe engineer shall direct." "The whole of the ironwork is to bbe properly freed from rust, and to have one coat of grey colouur before leaving the contractor's yard, and after completion i is to have two other coats of best oil colours, the whole of thhe paint to be of such character and quality as the enginee'er may approve." "Before any paint is applied, the ironworlrk must be clean and free from scales and rust. It shall havve two coats of thick red-lead paint approved by the engineer.i." "The exposed parts of all ironwork provided by the conn-tractor shall be scraped and painted with three coats of antiti-corrosive paint, approved by the engineer, well rubbed annd worked on the metal, no coat to be applied till the surface i is dry. The covered parts of all ironwork shall be scraped annd made clean, and have two coats of coal-tar put on hot." "All ironwork, before leaving the contractor's works, and afte:er inspection, to be well cleaned and painted one coat of goood oil colours, and after being fixed in place to be painted witlth two coats of best oil colours, the paint to be approved by thhe engineer." "The whole of the ironwork, before leaving thhe contractor's yard, is to be thoroughly cleaned and painte,ed one coat of good red-lead paint, and when the girders an,re erected complete to receive a second coat of good oil colour.r." "The bolts are to be heated to the temperature of melted lea,dd, and then dipped in boiling linseed oil." "The bolts and nut,ts are to be dipped twice when hot into hot boiled linseed oil immediately after they are made. The piles, bracings, and girders are to be protected by a good coat of tar or lime laid on hot in manner as may be directed by the engineer. All the *wrought*-iron sleepers, ties, clips, cotters, and other small parts, as may be directed by the engineer, must be heated and dipped while hot into boiled linseed oil. When dry all the small parts, with the exception of the tie-bars, must be properly packed in strong cases ready for shipment." "All ironwork shall be immersed while hot in a composition of coal-tar and asphalt in such manner as may be approved or directed by the engineer. After erection all the ironwork shall be carefully and evenly payed over with tar-asphaltum paint."

More detailed specifications require that all inaccessible surfaces must be well cleaned, painted or oiled, before the parts are assembled, and " the whole of the ironwork before it is riveted up, and, when practicable, after it is drilled, must be dipped in boiled linseed oil. It must be previously scraped clean, and, if required by the engineer, dipped in a weak solution to ensure a perfectly clean surface. It must receive two coats of good red-lead paint before being sent out of the works, the character and quality of the paint to be approved by the engineer." " After the parts of each girder and bridge have been inspected and approved, they are to be marked to their places in some manner to the satisfaction of the engineer. Each piece is then to be carefully cleaned from all rust, and then painted with one coat of red lead and linseed oil, and is afterwards to have one coat of good oil paint, proper time for drying being allowed between the application of each coat. In all cases the quality of any articles supplied must be the best of the class or rate ordered." " All plates, bars, angles, and T-irons, are to be dipped in hot boiled linseed oil while they are at a temperature of about 200° F. immediately after they are taken from the rolls and sheared, so as to prevent any corrosion of the surface from exposure to the atmosphere, and all the ironwork, before leaving the works, must have one coat of best anti-corrosive paint to be approved by the engineer, all damaged parts or fresh surfaces being first cleaned and re-coated. After erection, the ironwork shall be painted with two or three coats of good oil paint of quality and colour to be approved by the engineer. The engineer may order any tests to be made of the quality of the paint that he may consider advisable at the cost of the contractor."

" After the columns and girders are made, complete, and passed, the whole of the ironwork must receive two coats of the best metallic oxide paint of different colours, to be approved by the engineer or his inspector as directed."

A specification of Mr. Bouscaren, M. American Soc. C.E., is, " All iron and steel before leaving the shop shall have all loose scales scraped off, and shall be thoroughly coated with boiled linseed oil. All planed or turned surfaces shall be coated with white lead mixed with tallow. All inaccessible surfaces shall be painted before being put together with two coats of red lead or other metallic paint approved by the chief engineer. After erection, the entire structure shall be painted with two coats of the same paint. No painting shall be done in wet or freezing weather. All depressions in the erected structure where water is liable to collect, shall be drained by suitable drain holes or filled with approved waterproof mastic."

A specification for bridge-cylinder rings demands that, " the cast iron sections of the bridge cylinders after being cleaned, shall be heated to a temperature of 300° F. and immediately immersed in a bath of boiling coal-tar and asphaltum, five parts of coal-tar and one of asphaltum. Each section shall remain in the bath for a period to be directed by the engineer, but not less than fifteen minutes. The temperature of the bath shall be kept to the boiling point during the time the casting is immersed."

A stereotyped specification for painting iron or steel structures can hardly be drafted, for metallic structures are erected for so many different purposes, and are subject to corrosive influences of various kinds, some of little and others of special power, that ordinary provision against corrosion and fouling may be adequate or inadequate according to the special circumstances. A comprehensive perusal of the chapters of this book will enable a specification to be composed respecting the preparation of the surface to be coated, the character of the paint to be used, the method of application, &c. &c., that will suffice to secure, as far as practicable, any metallic structure from corrosive influences and fouling, which is of vital importance, for the commerce of the world and the lives and property of its inhabitants could not now be sustained or defended as they are without the existence and preservation of iron and steel ships and metallic structures.

Read the First Part of this volume, on ' The Safety and Preservation of Iron and Steel Structures,' for much information relating to corrosion and its prevention in bridges, ironwork in buildings, promenade and landing piers, rails, water pipes, and all metallic structures, whether submerged, buried in the earth, unsubmerged, partly, or occasionally submerged.

CHAPTER XIII.

SCAMPING TRICKS AND PAINTING.

THE following narrative, founded on fact, shows the manner in which scamping tricks have been carried out in painting. Two old sub-contractors met, and the following is what passed between them.

" Well, how are you ? "

" All right; I've been painting, and I'll tell you how II got 'extras' out of it. The specification said, 'the iron-work to have three coats of good oil paint.' It did not even say the metal was to be cleaned before the paint was spread, but we did it, more because we were afraid the stuff we put on would never hold on long if we did not clean the plates a bit, and that the paint might begin to crack and fall off before we were clear of the job altogether."

" You mean, left it and gone to another ? "

" Precisely. So we did rather more than the specification mentioned, although, perhaps, some of the other general clauses would have made it somewhat awkward for us. As for 'good oil paint,' have you ever seen anything else according to the sellers ? "

" Why, never ; of course, never. Everything is good that any one wants to sell, and I don't know it is not a libel to say it is not the best. Well, there happened to be a sale on at a builder's near the works, and I bought the things there, white lead, oil, turps, and driers, and they went cheap ; ; but I think the price was worth the goods. The white lead was streaky and hardly white, the oil was called ' best oil,' and that was enough for me, for the price suited ; the turpentine was equal to it ; and what the driers and colouring

matter were made of, any one that wants had better find out.
I mixed the things as I liked. My lad did it. It was a
funny kind of paint, but we used plenty of oil to make it
flow nice and easy, besides, we had to make it stick to the
metal, but when the oil is gone, it will be like whitewash,
anyhow it made it stick and look shiny. Don't ask me
whether it was anti-corrosive. It suited me well enough,
and ran in at about 1d. a lb."

" Oh gracious ! It must have been choice."

" It was to me, for it suited. You know, I believe a lot
of people never think it possible for rusting to go on *under*
the paint, but, when rust forms, it will go on, and increase
till entirely removed, and really anti-corrosive paint is put
on the dry, bare, pure metal surface, and not on the scale of
wrought iron or steel ; besides, the paint itself may cause
the metal to rust. I knew my man though, so just put it on
a bit thick, which I could afford to do, and made it shine
lovely."

" Fill your glass. Joe, bring the lads in, for I'm going to
tune up. Here is an old friend, and now I'll strike up, and
this is my latest. It is entitled :—

'THE FESTIVE PAINTER, HIS LITTLE GAME.'

" Three cheers for oil and lead I say,
They give me smoke and beer each day.
We buy it cheap to lay on thick,
And add some oil if 't will not stick,
Yes, rather.

" Don't ask of what the lead is made,
Or how much we, for it, have paid,
Or if there's acid in the oil.
You should not ask this son of toil.
No, rather.

" The driers, turps, and all the rest,
Of course are just the very best.
We mix them till we cannot stand,
And all the skin comes off our hand.
Yes, rather.

" We dab it on as it will go,
    And thin it out to make it flow.
    Yes ! on it goes, and on it must,
    For we don't care for dirt or rust.
                No, rather.
" It's nought to us if rust works through,
    In pits or flakes in year or two,
    Or paint falls off by night and day,
    That is if we are miles away.
                Yes, rather.
" We care not how the paint is made,
    So long as we are nicely paid,
    And it has got a lovely gloss,
    For that is sure to please the boss.
                Why, rather. Yes, rather. Good nighhit."

For descriptions of scamping tricks in other branchbhies
of engineering, see 'Scamping Tricks and Odd Knowleddjge
occasionally practised upon Public Works. Chronicled frcoom
the confessions of some old practitioners.' Published 1 lby
Messrs. E. & F. N. SPON, 125 Strand, London.

INDEX

TO

PARTS I. AND II.

———◦◦◦———

2 B

LONDON: PRINTED BY WILLIAM CLOWES AND SONS, LIMITED,
STAMFORD STREET AND CHARING CROSS.

Crown 8vo, cloth, 2s. 6d.

SCAMPING TRICKS

AND

ODD KNOWLEDGE

OCCASIONALLY PRACTISED UPON PUBLIC WORKS.

CHRONICLED FROM
THE CONFESSIONS OF SOME OLD PRACTITIONERS.

By JOHN NEWMAN,

Assoc. M. Inst. C. E., F. I. Inst.

REVIEWS OF THE PRESS.

ENGINEERING NEWS (New York).

"This readable and interesting book is arranged as a conversation between two old sub-contractors, in the course of which they deliver themselves of numerous yarns relating to methods practised on various kinds of works, to deceive the engineers and obtain the much-desired 'extras,' thus indicating some of the points to be especially looked after in superintending the construction of works. A still more interesting and valuable feature of the book, however, is that it is full of practical hints and notes upon different methods of carrying out different kinds of work under varying circumstances, giving also advice as to the merits of the different methods."

THE BRITISH ARCHITECT.

"We take the following story from a series of amusing narratives of 'Scamping Tricks and Odd Knowledge occasionally practised upon Public Works.'"

INDUSTRIES.

"This book is out of the run of ordinary professional works, inasmuch as it is intended, not so much for the purpose of showing how public works are to be carried out, as to point out some of the tricks which are practised by those who do not wish to carry them out properly, and to name some methods, founded on practical experience, adopted by sub-contractors and others to cheaply and quickly execute work.

"The young engineer or inspector will find many things in the book which will at least cause him to pay attention to special points in the different departments of civil engineering construction. Such matters as piles, which are chiefly hidden from view, seem to require careful inspection, and in fact all work which is covered up when the structure is completed."

INDIAN ENGINEERING.

"This is an entertaining little book. It abounds with stories of gross cheating. The ingenuity displayed in hiding the results of some of the frauds may be useful in setting young engineers on their guard against the over-plausible."

THE ENGINEERING REVIEW.

"The book should be read by all who are in charge of works of this kind."

THE ENGINEER AND IRON TRADES ADVERTISER (Scotland).

"The somewhat uncommon title of this book will in itself prove a ready attraction to the ordinary student of current literature. The title page alone is characterised by a curious vein of humour. The author has, however, a serious and a most important object in view.

"There is a peculiar charm in it not usually found in works where technical details require to be recorded. The many 'dodges' indulged in by these ideal contractors will come as 'eye-openers' to those unacquainted with the subject. We have no hesitation in saying that the volume before us is likely to serve a good purpose, and it is deserving of a wide circulation."

E. & F. N. SPON, 125 STRAND, LONDON.

NOTES ON CONCRETE AND WORKS IN CONCRETE.

By JOHN NEWMAN,

Assoc. M. Inst. C.E., F.I. Inst.

Second Edition, Revised and much Enlarged.

REVIEWS OF THE PRESS.

FIRST EDITION.

(Similar notices have appeared with respect to the Second Edition.)

ENGINEERING.

"*An epitome of the best practice, which may be relied upon not to mislead.*"
"The successful construction of works in concrete is a difficult matter to explain in books.
"All the points which open the way to bad work are carefully pointed out."

IRON.

"As numerous examples are cited of the use of concrete in public works, and details supplied, *the book will greatly assist engineers engaged upon such works.*"

THE BUILDER.

"A very practical little book, carefully compiled, and *one which all writers of specifications for concrete work would do well to peruse.*
"*The book contains reliable information for all engaged upon public works.*
"A perusal of Mr. Newman's valuable little handbook will point out the importance of a more careful investigation of the subject than is usually supposed to be necessary."

AMERICAN PRESS.

BUILDING.

"To accomplish so much in so limited a space, the subject-matter has been confined to chapters.
"*We take pleasure in saying that this is the most admirable and complete handbook on concretes for engineers of which we have knowledge.*"

E. & F. N. SPON, 125 STRAND, LONDON.

EARTHWORK SLIPS AND SUBSIDENCES UPON PUBLIC WORKS.

By JOHN NEWMAN, Assoc. M. Inst. C.E., F.I. Inst.

REVIEWS OF THE PRESS.

Engineering News (New York).

" The book is of a practical character, giving the reasons for slips in various materials, and the methods of preventing them, or of making repairs and preventing further slips after they have once occurred. The subject is treated comprehensively, and contains many notes of practical value, the result of twenty-five years' experience."

The Builder.

" We gladly welcome Mr Newman's book on slips in earthworks as an important contribution to a right comprehension of such matters.

" There is much in this book that will certainly guard designers of engineering works against probable, if not against possible, slips in earthworks.

" The capital cost of a work and the cost of its maintenance may both be very sensibly reduced by attention to all the points alluded to by the author.

" We are glad to see that the author enters at some length into the subject of the due provision of drainage at the backs of retaining walls, a matter so often neglected or overlooked, and carries this subject to a far larger one, the causes which tend to disturb the repose of dock walls. His remarks on these matters are well worthy of consideration, and are thoroughly practical, and the items which have to be taken into account in the necessary statical calculations very well introduced.

" In conclusion, we may say that there is plenty of good useful information to be obtained from this work, which touches a subject possessing an exceedingly scanty vocabulary.

" It contains an immense deal of matter which must be swallowed sooner or later by every one who desires to be a good engineer."

Building News.

" Mr. John Newman, Assoc. M. Inst. C.E., has written a volume on a subject that has hitherto only been treated of cursorily.

" Useful advice is given, which the railway engineer and earthwork contractor may profit by.

" The book contains a fund of useful information."

Builders' Reporter and Engineering Times.

" The book which Mr. John Newman has written imparts a new interest to earthworks. It is, in fact, a sort of pathological treatise, and as such

E. & F. N. SPON, 125 STRAND, LONDON.

may be said to be unique among books on construction, for in them failures are rarely recognised. Now in Mr. Newman's volume the majority of the pages relate to failures, and from them the reader infers how they are to be avoided, and thus to form earthworks that will endure longer than those which are executed without much regard to risks.

" The manner of dealing with the subsidences when they occur, as well as providing against them, will be found described in the book.

" It can be said that the subject is thoroughly investigated, and contractors as well as engineers can learn much from Mr. Newman's book."

NOTES ON CYLINDER BRIDGE PIERS,

AND THE

WELL SYSTEM OF FOUNDATIONS.

With which, by permission, is incorporated the Miller Prize (Institution of Civil Engineers) Paper on the same subject.

By JOHN NEWMAN,

Assoc. M. Inst. C.E., F.I. Inst.

REVIEW OF THE PRESS.

(Other reviews are to similar effect.)

THE BUILDER.

" Most of the chief points requiring attention in the design, sinking, or erection of cylinder piers or wells, either by compressed air, dredging, or open trench work, have been dealt with by the author; and upon a perusal of the eighteen chapters which form the volume, we find a good deal of practical information.

" The author has taken this opportunity to add numerous hints, alike useful to the resident engineer, bridge-builder, contractor, and last, though not least, the student.

" There is much to be learnt from the study of Mr. Newman's book relating to the construction of bridge, quay, dock, weir, and river-wall foundations."

E. & F. N. SPON, 125 STRAND, LONDON.